Design Paradigms

Design Paradigms
Case Histories of Error and Judgment in Engineering

HENRY PETROSKI
Duke University

CAMBRIDGE
UNIVERSITY PRESS

Published by the Press Syndicate of the University of Cambridge
The Pitt Building, Trumpington Street, Cambridge CB2 1RP
40 West 20th Street, New York, NY 10011–4211, USA
10 Stamford Road, Oakleigh, Melbourne 3166, Australia

First published 1994
Reprinted 1994, 1995, 1996

Printed in the United States of America

Library of Congress Cataloging-in-Publication Data
Petroski, Henry.
Design paradigms : case histories of error and judgment in
engineering / Henry Petroski.
p.   cm.
Includes bibliographical references and index.
ISBN 0-521-46108-1 (hc). – ISBN 0-521-46649-0 (pbk.)
1. Engineering design – Case studies.   2. System failures
(Engineering) – Case studies.   I. Title.
TA174.P473   1994
620'.0042 – dc20   93-32560
CIP

A catalog record for this book is available from the British Library.

ISBN 0-521-46108-1 hardback
ISBN 0-521-46649-0 paperback

To
William Petroski, P.E.,
brother engineer

Contents

Preface

Possibly the greatest tragedy underlying design errors and the resultant failures is that many of them do indeed seem to be avoidable, yet one of the potentially most effective means of improving reliability in engineering appears to be the most neglected. Historical case studies contain a wealth of wisdom about the nature of design and the engineering method, but they are largely absent from engineering curricula, perhaps because the state of the art always seems so clearly advanced beyond that of decades, let alone centuries or millennia, past. However, the state of the art is often only a superficial manifestation, arrived at principally through analytical and calculational tools, of what is understood about the substance and behavior of the products of engineering. Anyone who doubts this assertion need only look to the design errors and failures that occur in the climate of confidence, if not hubris, known as the state of the art.

The fundamental nature of engineering design transcends the state of the art. Thus it follows that historical case studies that illuminate those aspects of conceptualization, judgment, and error that are timeless constants of the design process can be as important and valuable for understanding technology and its objects as are the cal-

culus or the latest computer software. The lessons of the past are
not only brimming with caveats about what mistakes should not be
repeated but also are full of models of good engineering judgment.

This book argues for a more pervasive use of historical case studies in the engineering curriculum. There are several direct benefits
and cultural byproducts to be derived from such use: the patterns
of success and failure that have manifested themselves in engineering
for centuries are made explicit and readily available for extrapolation;
the lessons learned by our professional ancestors can be kept high
in the consciousness of succeeding engineering generations; an
awareness of the timeless elements of the engineering method and
their commonality across ostensibly disparate specialties can give a
theoretical foundation to the engineering curriculum that students
often find wanting; a common store of case histories and anecdotes
can serve as a lingua franca among the various engineering fields
that will emphasize their similarities; and, finally, engineering students can be made more aware of the roots of their profession and
its relationship to society.

The objective of this book is not only to present a model for
explaining how errors are introduced into the design process but
also to provide a means by which practicing designers may avoid
making similar errors in their own designs. The use of pithy and
classic anecdotes set in familiar design situations is an excellent
means for abstracting general principles while at the same time providing unifying themes and useful lessons that will be remembered.
But it would not serve the intention of this work merely to relate a
long list of familiar and hackneyed stories about classic engineering
failures, such as the Leaning Tower of Pisa or the Tacoma Narrows
Bridge. Already well-known case studies cannot in themselves necessarily help us to reach new conclusions about design or to eliminate
errors in design generally.

What can both teach us about design and improve our practice
of it is a carefully selected group of case histories or studies that
illuminate different, even if often overlapping, aspects of the design

process and that thus serve as paradigms for both theory and practice. To be effective as a paradigm of error, a particular case study not only must be capable of being presented as a fresh and memorable story (or at least a familiar story told from a fresh perspective) but also must be capable of evoking a host of related case studies or horror stories in a wide variety of engineering contexts and disciplines, thus demonstrating how over time the same or similar mistakes have led to repeated failures of design. If a paradigmatic case study can do this, then it is likely to embody a general principle of design error that can also arise in new design situations. Thus the paradigm will have the dual value of providing a guide for understanding the design process as well as presenting a means of improving it by alerting the designer to common pitfalls in design logic.

The anecdotal nature of the paradigms presented in this book is intended to evoke associations with the real situations in which human designers necessarily find themselves every day. The guiding principle has been to select, explicate, and validate a group of paradigms that illustrate forcefully how certain types of errors can be introduced into or avoided in various aspects of the design process. However, this collection of paradigms is not intended to constitute a unique, distinct, exhaustive, or definitive classification of design errors, but rather to show the efficacy of the approach.

Some more recent examples of engineering failures have been included to emphasize that the construction of a historically based paradigm is not just a scholarly exercise, but one that is relevant today and will remain relevant for avoiding error in engineering designs of the twenty-first century. If persuasive paradigms and instructive case histories can be multiplied and disseminated in an effective way, there is reason to believe that they can become as important a part of the designer's intellectual tool kit as are laws of mechanics, rules of thumb, and computer models.

This book is intended not only for engineering students; practicing engineers may find it a different perspective on their profession, and general readers interested in gaining some insight into the engineering method might find it a painless introduction to the subject. Since the book's concrete case histories and lessons are intended to be accessible to engineering students and practitioners from all fields and at all levels, as well as to the general reader, jargon has been eschewed and equations have been all but eliminated. Thus the reader need bring no prerequisites to the book other than a mind open to the idea that what engineers were doing in times past has some relevance to what engineers do today.

The work leading up to this book was made possible in part by a grant from the National Science Foundation, for which I am grateful. Several of the chapters have been adapted from refereed articles, which I have published in the following journals: *Journal of Engineering Design* (Chapter 1), *Research in Engineering Design* (Chapter 2), *Journal of Mechanical Design* (Chapter 3), *Structural Safety* (Chapter 4), *Civil Engineering Systems* (Chapter 5), *Structural Engineering Review* (Chapter 7), and *Journal of Performance of Constructed Facilities* (Chapter 8). Much of Chapter 6 is scheduled to appear in *Civil Engineering*, and most of the ideas in Chapter 10 appeared first in my engineering column in *American Scientist*. Helpful comments of anonymous reviewers and the editors of these publications have often made these chapters more focused, but any errors that remain are naturally artifacts of my own humanity.

This manuscript, like most of my recent work, was first drafted in the carrel kindly assigned to me by Albert Nelius in the William R. Perkins Library of Duke University. While I am, as always, indebted to many librarians, both at Duke and throughout the interlibrary loan network, I must once again single out for special acknowledgment and praise Eric Smith of Duke's Aleksandar S. Vesić Engineering Library. His knowledge and help have been indispensable. As always, my wife, Catherine, has understood my need to retreat to my study at home and remains my fairest reader.

· · ·

I

Introduction

The concept of failure is central to the design process, and it is by thinking in terms of obviating failure that successful designs are achieved. It has long been practically a truism among practicing engineers and designers that we learn much more from failures than from successes. Indeed, the history of engineering is full of examples of dramatic failures that were once considered confident extrapolations of successful designs; it was the failures that ultimately revealed the latent flaws in design logic that were initially masked by large factors of safety and a design conservatism that became relaxed with time.

Design studies that concentrate only on how successful designs are produced can thus miss some fundamental aspects of the design process, which is difficult enough to articulate as it is. Yet while practicing designers especially are notorious for saying little, if not for consciously avoiding any discussion at all of their own methodology, there have been some notable exceptions (especially, e.g., Glegg, 1973, 1981; Leonhardt, 1984), and when these engineers have reflected on the design process, they have acknowledged the important role that failure plays in it. Although often an implicit and tacit part of the methodology of design, failure considerations and proactive failure analysis are essential for achieving success. And it is precisely when such considerations and analyses are incorrect or incomplete that design errors are introduced and actual failures occur. Understanding how errors are made and can be avoided in the design process can help eliminate them and can illuminate the very process of design.

The role of failure in successful design has been argued explicitly in the context of structural engineering (cf. Petroski, 1985), but the argument applies by analogy in fields as seemingly divergent as aerospace, chemical, computer, electrical, and mechanical engineering. Alexander (1964) has reasoned that whenever any designer does not merely copy exactly what has already been made and pronounced successful – that is, acceptable – then it is difficult to say whether a new or modified design will be or will continue to be successful. A building or a bridge may be declared successful during decades of problem-free service, but if it suddenly collapses we may find that a serious design flaw was present all along. Thus, the elevated walkways in the Kansas City Hyatt Regency Hotel were considered safe enough to hold the crowds that they did until they collapsed on the afternoon of Friday July 17, 1981; and the Mianus River Bridge near Greenwich, Connecticut, daily carried the heavy traffic of Interstate 95 until it suddenly collapsed early in the morning of June 28, 1983.

Recognizing failure is something "even the simplest" person can do, and Alexander has noted that even if only a few of us have "sufficient interpretative ability to invent form of any clarity, we are all able to criticize existing forms." He considers this an "obvious point," acknowledging that it is at least as old as Pericles, who is quoted as saying, "Although only a few may originate a policy, we are all able to judge it." So when a structure collapses or a design fails to live up to its promise, the lesson should be accessible to all, and even a jury of laypersons is presumed capable of ruling on culpability.

While it may be easy to recognize failure after the fact, engineers are expected to anticipate and obviate it in their designs. The idea of proactive failure analysis has been elaborated upon by Alexander: he gives some forceful examples of how much more natural it is for us to recognize failure or "misfit" between context and form than to recognize the success or fitness of a design in meeting stated requirements. Indeed, Alexander goes so far as to say:

We are never capable of stating a design problem except in terms of the errors we have observed in past solutions to past problems. Even if we try to design something for an entirely new purpose that has never been conceived before, the best we can do in stating the problem is to anticipate how it might possibly go wrong by scanning mentally all the ways in which other things have gone wrong in the past.

According to Lev Zetlin (1988), whose career as a prominent structural design engineer included many major projects:

Engineers should be slightly paranoiac during the design stage. They should consider and imagine that the *impossible could* happen. They should not be complacent and secure in the mere realization that if all the requirements of the design handbooks and manuals have been satisfied, the structure will be safe and sound.

Furthermore, according to Zetlin (as quoted in Browne, 1983), what is needed to improve designs and the reliability of structures is "preventative and curative engineering." He elaborated on what he meant by explaining his own design and review technique: "I look at everything and try to imagine disaster. I am always scared. Imagination and fear are among the best engineering tools for preventing tragedy." No one is advocating that today's engineers become immobilized by the design process, but Zetlin merely echoed what great engineers like Robert Stephenson and Herbert Hoover also reported: they lost sleep over the design problems they took to bed with them. Rather than being the curse of engineering, this manifestation of concern should be worn by the profession as a badge of honor.

Other reflective engineers have made similar observations, and in order to imagine what can go wrong with one's current design, it is clearly advantageous to know what has gone wrong with others in the past. The dictum of Santayana that those who do not remember the past are condemned to repeat it should always weigh heavily on the minds of engineers. And it would seem to be especially important that we think more about past design failures in light of the report of a recent National Science Foundation workshop group on occurrence of errors that "in many cases the same errors are repeated again and again" (Nowak, 1986).

Because of their long history and their scale, large civil engineering structures and mechanical engineering systems have had their designs and failures most thoroughly documented, and thus there are archives of information about them that are unequaled. Furthermore, because they are so visible and public, the record of failures of civil structures and large public systems can be expected to be more complete and wide-ranging than those of other failures, with technical reports often being supplemented by contemporary nontechnical reports that place the failures in the broader social and human context in which designers must necessarily operate and in which errors occur.

Although structural engineering examples are often the most common in arguing for the role of failure in the design process, this concentration by field in no way restricts the applicability of the general concepts involved. Publications in computer, electrical, manufacturing, mechanical, and other fields of engineering frequently contain reports and analyses of failures that at first might seem far afield.

The relevance of lessons learned in structural engineering for other engineering applications was emphasized by the publication of a long interview with a bridge engineer in a journal of the Association for Computing Machinery (Spector and Gifford, 1986), an introduction to which declared: "Though some computer systems are more complex than even the largest bridges, there is a wealth of experience and insight in the older discipline that can be of use to computer systems designers, particularly in such areas as specification, standardization, and reliability." The editors clearly felt that there was much to be learned by analogy from a field with a long history by one with a very short one. Interestingly, bridge failures played a prominent role in the interview.

This view of the generality of lessons to be learned from failures was reiterated in a special report on managing risk in large complex systems published even more recently in *Spectrum,* the magazine of the Institute of Electrical and Electronics Engineers (Bell, 1989). According to the report's introduction:

Although some of these case studies examine systems that are neither electrical nor electronic, they highlight crucial design or management practices pertinent to any large system and teach all engineers important lessons. What large systems have in common counts for more than how they differ in design and intention.

Aeronautical and aerospace engineering failures have also been the subject of much professional attention, not to mention extensive coverage in the mass media. Nuclear engineering incidents like the 1979 loss of coolant at the Three Mile Island plant near Harrisburg, Pennsylvania, and the 1986 explosion and fire at the Chernobyl plant near Kiev, which released radioactive material that spread well beyond its origin in the Soviet Union, have become shibboleths for discussions of the risks of modern high technology. Chemical engineering shortcomings have been the focus in discussions of industrial accidents such as the 1984 release of toxic gas from a Union Carbide insecticide plant at Bhopal, India, which killed more than 2,000 people and harmed about 150,000 more. Mechanical engineering failures can result in death due to exploding gasoline tanks in automobiles and trucks, not to mention less life-threatening but wide-reaching effects such as massive recall campaigns and product liability suits. And electrical and software engineering have had to deal with questions of reliability and assurance in the wake of massive power blackouts and telephone system breakdowns that have inconvenienced tens of millions of people at a time. In short, no field of modern engineering is untouched by the effects of failures and their invaluable lessons.

It appears incontrovertible that understanding failure plays a key role in error-free design of all kinds, and that indeed all successful design is the proper and complete anticipation of what can go wrong. It follows, therefore, that having as wide and deep an acquaintance as possible with past failures should be at least desirable, if not required, of all engineers engaged in design. Understanding from case histories how and why errors were made in the past cannot but help eliminate errors in future designs. And the more case histories a designer is familiar with or the more general the lessons he or she can draw from the cases, the more likely are patterns of erroneous thinking to be recognized and generalizations reached about what to avoid.

Human Error

Many thoughtful researchers have reflected on the sources of error in design, including Blockley (1980), who has stated flatly that in the final analysis "all error is human error, because it is people who have to decide what to do; it is people who have to decide how it should be done; and it is people who have to do it" (cf. FitzSimons, 1988). Some might argue with Blockley's singular classification of error, but the human element clearly increases failure rates and thus reduces the reliability of our designed artifacts and systems compared to analytical predictions that obviously ignore or underestimate the human element in design.

According to Ingles (1979), statistical methods for quantifying variability in design are insufficient if the human sources and nature of variability are not more closely examined. Writing more recently, Santamarina and Chameau (1989) have reiterated the problem, and have reemphasized the need to look to the human element in design methods and practices. Nowak and Tabsh (1988) have stated that the "major challenge to reliability theory was recognized when the theoretical probabilities of failure were compared with actual rates of failure [and] actual rates exceed the theoretical values by a factor of 10 or 100 or even more." They identified the main reason for the discrepancy to be that the theory of reliability employed did not consider the effect of human error. A host of further studies, such as referenced in the work of Blockley (1980), Ingles (1979), and others, make it clear that there is a strong sense that improved reliability in design will come only when our already highly developed analytical, numerical, and computational design tools are supplemented with improved design-thinking skills. While artificial intelligence and expert systems have been promised as solutions to the problem of human error, the design of computer-based methods will itself benefit from an understanding of human error and how to reduce it.

We should expect to learn more about human error in design from case studies of actual design errors and failures. However, as Blockley (1980) has noted in the context of structural design, the commercial nature of engineering works against the wide dissemination of accounts of human error, except in cases of failures that lead to some form of public inquiry. According to Blockley, although we have accounts of successful projects, "from these there is the least to learn." The gravity of the situation was further emphasized during hearings before the U.S. House of Representatives Committee on Science and Technology (1984) in which the practice of sealing the testimony of legal proceedings dealing with liability over design failures was deplored. Such a practice deprives the larger design community of invaluable lessons learned.

Human error in anticipating failure continues to be the single most important factor in keeping the reliability of engineering designs from achieving the theoretically high levels made possible by modern methods of analysis and materials. This is due in part to a de-emphasis on engineering experience and judgment in the light of increasingly sophisticated numerical and analytical techniques. According to the report of a working group on the occurrence of errors (Nowak, 1986), some observers hold that society has found the current average level of risk acceptable, while others note that increased legal liability costs show that society has not. Whatever society's perception, since human beings are the only feature common to the diverse systems surveyed in a variety of historical, geographical, and economic contexts, the discrepancies between predictions and reality have been attributed primarily to the human element in the design, construction, and maintenance processes. Education, motivation, and quality control provide clear, if not necessarily easily implemented, ways of reducing human error in manufacturing, construction, and maintenance, but ways to eliminate human error from the design process are much less obvious.

We now have very sophisticated theories of structures and elaborate multipurpose computer programs that are capable of quite refined analysis, but these have not led to an improvement in the reliability of engineering design. Indeed, more than one survey of engineering failures have concluded that refined methods of analysis would not prevent future failures. The foundation engineer Peck (1981), speaking of dams, concluded that "nine out of ten recent failures occurred not because of inadequacies in the state of the art, but because of oversights that could and should have been avoided." He pointed out that the "problems are essentially nonquantitative" and the "solutions are essentially non-numerical." Peck acknowledged that improvements in analysis and testing might be profitable, but felt it was also likely that "the concentration of effort along these lines may dilute the effort that could be expended in investigating the factors entering into the causes of failure." Hauser (1979), after reviewing a survey of about 800 European failures, concluded that "the most efficient way to improve structural safety or to reduce the overall effort to maintain a certain level of structural safety is to refine the methods of data checking [to catch design errors] and not to refine the models of analysis." There has been little change of concern in the ensuing decade (cf. Santamarina and Chameau, 1989), principally because failures continue to occur at unexpectedly high rates. Indeed, such a state of affairs and the accompanying questions of liability led the American Society of Civil Engineers in 1988 to draft a manual of professional practice, *Quality in the Constructed Project*.

It is inevitable that errors are going to be made in design. Some conceptual designs are just bad ideas from the start, and it is the self-critical faculty of the designer that must be called into play to check him- or herself and to abandon bad ideas on the drawing board (cf. Petroski, 1985). Other designs are fundamentally sound conceptually, but they can be weakened by poor choices of components or inferior detailing or by seemingly simple but poorly considered design changes. It is the function of checking, whether by the original designer or by peers, to catch the omission of a critical calculation, the lapse in logic, the error in analysis, or the mistake in mathematics. But all too often the process of checking – an integral component of the design process itself – is myopic. The original designer can continue to overlook the same errors of commission or omission, and the peer can nod at the faulty logic (cf. Stewart and Melchers, 1989).

Alexander (1964) has reminded us of the fundamental fact that the main objective of design and the study of its methodology is to make better designs – that is, to improve reliability. Looking for rules of success in the design process as it is currently practiced will not necessarily accomplish this goal, for it is design as practiced that is failing to live up to its desired or theoretical potential. Rather, we must look anew at the design process and learn how better to identify the causes of error in it. Since the ideally successful design properly anticipates all relevant and possible ways in which failure can occur, it is imperative to understand how failure is introduced by human designers into the design process.

The Case for Historic Case Studies

Even when information about contemporary failures is available, there can be honest disagreement among experts over the ultimate cause of a failure, in part because of human nature and our reluctance to be perfectly candid about our own errors. Thus, even when modern case studies are available, they may provide a skewed perspective on the actual design process because of pending law suits, because of professional reputations that are at stake, or because of commitments to current theories. Such difficulties and complications argue for looking to historical case studies for examples of human error and how to deal with it. Although new theoretical and computational design tools have made the old obsolete, the nature of the design problem and the design logic and thought processes used to solve it have remained essentially unchanged from ancient times.

Design today is commonly done in collaboration, and thus all the elements of the process are not necessarily embodied in a single individual. Mainstone (in Pugsley, Mainstone, and Sutherland, 1974) has recognized this difficulty and has written that he has "found it easier in some ways to get under the skin of earlier designers . . . than to do the same in the case of the more typical design teams of today." Studying earlier designers can thus complement studies of design groups.

Whenever the most gifted thinkers and designers, whether classical or modern, have without external prompting reflected on their art and its failings, they have given us potentially very valuable data on the nature of engineering design that applies beyond the specific case study. Since all design necessarily must conform to both technical and nontechnical constraints, the most meaningful data on the design process tend to be those which place a given design in a contemporary context. And since failures contain more unambiguous information than successes, the most fruitful data that any designer can be provided are case studies of failure or the explicit avoidance of failure.

Classic and historical case studies have the further advantage that they are embedded in a more or less static and closed context, and thus they may be considered virtually timeless and unchanging examples. Recent case studies have the disadvantage that they are often the focus of ongoing litigation or investigation, thus not only making them subject to considerable secrecy and adversarial claims, if not downright misrepresentation, but consequently also making their interpretation subject to change as new information becomes available. Although now-historical case studies were no doubt subject to the same human problems in their own times, and although they are also subject to revisionist readings of history, they may generally be considered among the most objective and natural sources of data on the design process that are available.

Since no one today should have a vested interest in the outcome of an investigation into design errors made a century or more ago, historical case studies provide the raw data for controlled experiments and investigations into the characteristics of human errors and how to avoid them in modern design. The perspective of time enables everyone to participate on an equal basis in the interpretation and use of the historic record. Cogent case histories can be exploited for the dual purposes of explaining how error is introduced into and avoided during the design process itself and of providing a means for eliminating error from future designs.

For all of its use of sophisticated mathematics and computational models, design today involves the human mind in fundamentally the same way it did for the first builders. This is implicit in Ferguson's (1977) classic description of the nonverbal aspect of conceptual design (namely, that it relies more on pictures than on words or numbers), and it is evident in all current attempts to model the design process and its methodology in artificial intelligence and expert systems (cf. Ferguson, 1992). Although we freely use the term "engineering method," its precise definition is curiously elusive and is yet to be articulated in a universally agreed-upon form. But for all its fuzziness, the engineering method is no less practiced than is tying our shoes in the absence of directions on a package of laces. Lessons that are seemingly obsolete and seemingly as simple and self-evident as tying a bow can provide real insight into some of the most fundamental aspects of engineering design and its method.

Vitruvius, writing of the state of the art in the first century B.C.E., related some sketchy case histories of failures, including problems to be avoided in building walls and the story of one Paconius, an engineer-contractor hired to move a heavy marble pedestal. Evidently, in order to get the job, Paconius departed from accepted practice and devised a novel way of encasing the squarish monolith in a spool-like cage of wood over which a hauling rope was wound. But his scheme failed, and Vitruvius's lesson in relating the case history of Paconius's failure appears at first to be merely an argument against departing from tradition. By doing things the way they have always been done, one does not risk failure. But there are much more subtle lessons to be drawn from Vitruvius's story of failed schemes, and these are elaborated upon in Chapters 2 and 3.

Galileo, whose seventeenth-century *Dialogues Concerning Two New Sciences* includes what is considered the first attempt to provide an analytical basis for the design of beams to carry designated loads, also recognized the responsibility of the human agent in making things correctly – or incorrectly. Galileo's seminal treatment of the strength of materials was motivated by some stories of structural failures that were inexplicable within the then-current state of the art of shipbuilding and construction practice. Because Renaissance engineers did not fully understand the principles upon which they were making and building things, they committed the human errors that were the ultimate causes of the failures, of course. But rather than suggest that there should be no attempt to extrapolate from proven designs, Galileo set out to lay the foundations for a new engineering science that would give future engineers the mental and analytical tools to eliminate error from their conceptions and extrapolations. These points are elaborated upon in Chapters 4 and 5.

The modern tradition of using failures to expose ignorance rather than to stifle innovation thus has a model in Galileo. By the nineteenth century, when his rudimentary efforts at assessing the strength of materials had developed into the theories of beams and structures that enabled the railroads to expand their lines over great iron bridges, no less an engineer than Robert Stephenson was arguing in Britain for the publication of case histories of failures. "Nothing was so instructive to the younger Members of the Profession, as records of accidents," Stephenson wrote, and "older Engineers derived their most useful store of experience from the observations of those casualties which had occurred to their own and to other works, and it was most important that they should be faithfully recorded in the archives" (quoted in Whyte, 1975).

Stephenson was speaking from experience, for he had learned from the 1847 failure of his own Dee Bridge at Chester in northwest England (described in Chapter 6) to proceed with extreme caution when designs departed from common practice. He was also led by the numerous failures of early suspension bridges to come up with a brilliant, if expensive, alternative in his Britannia tubular bridge over the Menai Strait in northwest Wales (Chapter 7). But John Roebling, working at about the same time in the United States, came to a different conclusion after studying the same case histories (Chapter 8). Rather than being led to abandon the economical suspension bridge principle, Roebling learned from past failures what to design against and what mistakes to avoid. He used conceptual failure analysis to eliminate from his design for the Niagara Gorge Bridge, between western New York State and Canada, the weaknesses that were the downfall of the failed bridges. And what he was able to learn from failure was sufficiently profound for him eventually to double the length of the Niagara Gorge Bridge in the design for his masterpiece, the Brooklyn Bridge, realized posthumously in 1883.

In the half-century following the accomplishment of the Brooklyn Bridge, suspension bridge design evolved in a climate of success and selective historical amnesia (as described in Chapter 9) to the Tacoma Narrows Bridge. Now, a half-century after that bridge's co-lossal collapse, there is reason to be concerned that the design of newer bridge types will be carried out without regard to apparent cycles of success and failure. As shown in Chapter 10, the historical record of bridge failures contains some disturbingly regular patterns that, if not recognized and heeded, may be inadvertently continued. A familiarity with the details of past mistakes gained through the kinds of case studies presented here may be the surest way to break established patterns of failure and ensure more successful and reliable designs in the future.

2

Paconius and the Pedestal
for Apollo

A Paradigm of Error in
Conceptual Design

Errors can occur at all stages of the design process, but fundamental errors made at the conceptual design stage can be among the most elusive. Indeed, such errors tend to manifest themselves only when a prototype is tested, often with wholly unexpected or disastrous results. These, more than any other design errors, are invariably human errors, because a conceptual design comes only out of the uniquely human creative act of transforming some private concept from the designer's mind to some public concept that can be described to and modified by other humans. The creative act of conceptual design is the result of nonverbal thought (Ferguson, 1977, 1992), and as such often takes its initial public form as a sketch or drawing. If the concept is fundamentally flawed, this may be recognized intuitively at the instant the design is articulated in the mind or in words or drawings on paper, and so the concept can be rejected outright. But if the concept has a basic error that goes undetected, then the error tends to be all the more difficult to catch as the design progresses through evolutionary stages of modification and detailed design.

Whatever classification schemes might be employed or whatever distinctions might be drawn among identifiably different stages of the design process, the creative act of conceptual design is as old as civilization itself, for it can be argued that without such design there would be no civilization. Furthermore, because in its most fundamental stage conceptual design involves no overt modern theoretical or analytical component, there is no reason to believe that there is any essential difference in the way our most ancient ancestors conceived and we still do conceive designs. Nor is there any reason to believe that there is any essential difference in the ways in which our ancestors erred and we can err in our conception and thus inadvertently pursue flawed designs to fruition and failure.

The Ten Books on Architecture, written by the Roman Vitruvius in the first century B.C.E., documents the state of the art of building in ancient Greece and Rome and is generally considered the oldest surviving book on engineering. In his work, Vitruvius relates a case study of a design failure that may be taken as a paradigm of error in conceptual design. It is a story of how the Roman contractor Paconius went bankrupt moving a massive stone pedestal by making what seemed to him only slight cost-saving modifications to a classical method of moving heavy column shafts and architraves, the rectangular pieces of stone that spanned the distance between the tops of columns in ancient temples and other structures.

The evolution of designs has always been driven by economic, aesthetic, and other nontechnical functional reasons, and the story of Paconius and the pedestal brings out in an especially clear manner how easily a designer concentrating on satisfying one kind of objective in detail can overlook modifications that in retrospect prove to be disastrous to the functioning of the machine, structure, or process that is the primary purpose of the design.

In this chapter, along with the story of Paconius's error, a justification of the design changes made by Paconius will be offered in a context that shows how they could have appeared to be rational and wise at the time. It will also be shown how in fact the design changes could have been driven by reasoning that appeared to be removing objections to prior designs and eliminating ways in which in their own way the earlier designs were perceived to have failed to meet certain design criteria. After a discussion of how Paconius could have rationalized and defended his concept to his contemporaries, a failure analysis of the design will be carried out and lessons drawn. Collectively, the story of Paconius's ancient design and its rationalization and failure analysis constitute a paradigm that illuminates the conceptual design process. Understanding this paradigm has the potential for eliminating error in today's and future designs.

Some Background to the Story of Paconius

Few designs spring up from nothing, and so the story of how Paconius came upon his scheme to move a large stone pedestal from its quarry to the temple of Apollo at Selinous is best begun with some background on why and how large stones came to be moved the way they were in the ancient world. The basic design problem was, even back then, an old one: to transport without breaking massive pieces of stone to be used for monumental column shafts, architraves, pedestals, and the like. There often would be naturally occurring obstacles between quarry and building site, including bodies of water and uneven and undeveloped terrain, as well as developed regions that included narrow roads and restricted passageways. Because the stones were massive, when wagons were used they had to be very strong and massive themselves, and their wheels had to be prevented from getting stuck in soft ground. The various schemes to overcome such obstacles, often described and generally well known, include the use of barges, rollers, sledges, and the like. However, in certain circumstances ingenious methods were developed that appeared to have distinct advantages over the more familiar techniques. The stories of how these various methods evolved are classic stories of design and redesign, and they should impress upon modern-day designers and students of design the timelessness of the process. Although our problems are different, our techniques more sophisticated, and our physical and analytical tools more refined, underlying all the superficial advances in the conception and realization of artifacts are the same fundamental characteristics of design.

According to Coulton (1977), the most likely normal means of transporting large blocks of stone across horizontal distances was four-wheeled carts and wagons drawn by oxen (cf. Burford, 1960). Very heavy blocks taxed the strength of axles, however, and the four wheels placed considerable pressure on the ground or road surface, thus possibly giving rise to the design of wagons with six or eight wheels. Such improved wagons presented new problems, such as how to distribute the load effectively over redundant axles and how to steer the awkward vehicle. Sledges presented one means of carrying great weights over soft ground, and an alternative to both the sledge and more wheels was, of course, wider wheels. The principle of such a solution seems to have been introduced as early as twenty-five centuries ago in the construction of the temple of Diana at Ephesus, one of the first colossal Greek temples.

Vitruvius relates how Chersiphron, who, "not trusting to carts, lest their wheels should be engulfed on account of the great weights of the load and the softness of the roads in the plain" between the quarry and the temple site, conceived of a new plan. As with all new plans, Chersiphron's concentrated on and modified those features of an old one that threatened its success or caused it to fail. He saw that a way to prevent wheels from sinking easily into the plain was to make them wider. Further, if wider wheels were better, then the widest wheels might be best. Still further, if wagon axles would still be strained toward their limits, then eliminating axles would certainly obviate axles breaking. He achieved his objective by effectively eliminating the wagon with its wheels and axles altogether. Rather than load the circular cylindrical shafts of the great columns onto wagons, Chersiphron used each shaft itself as a wide wheel or roller. He used lead and iron to fit pivots into the cut-out center of the stone on each of its flat ends, and he fit a wooden frame around the stone so that it could be pulled along the ground like a giant roller.

Figure 2.1. Chersiphron's scheme for transporting circular columns (Larsen, 1969).

Not only did the column not get bogged down so easily in the soft ground, but also it was easier to pull and thus required fewer oxen. The scheme is shown in Figure 2.1.

After a successful scheme had been devised to transport column shafts to the temple site, it became desirable to transport architraves in an equally effective way. Since these stones were square or rectangular in cross section, they could not serve as wheels directly. However, Chersiphron's son, Metagenes, came up with a clever modification. He used the stone itself as a great axle and built wide wooden wheels, twelve feet in diameter, around either end of it, into which he fitted pivot cups that could receive the pins attached to a timber frame of the kind used by his father. The scheme, illustrated in Figure 2.2, was successful, and it had the added advantage of being able to transport column shafts that were tapered and would have been impossible to move by Chersiphron's method. Chersiphron and Metagenes published a book on their innovative designs, and it evidently served as the source of Vitruvius's familiarity with the story.

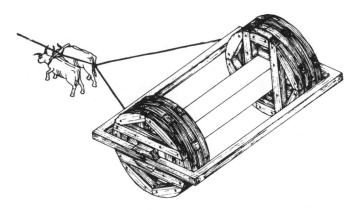

Figure 2.2. Metagenes' scheme for transporting architraves (Coulton, 1977; by permission of Cornell University Press).

The Scheme of Paconius: Its Genesis and Advantages

Paconius, a near-contemporary of Vitruvius, also seems to have read or at least heard about the new methods of transporting great pieces of stone, and he proposed to modify the latest scheme of Metagenes to move a large piece of stone (twelve by eight by six feet) needed to replace the cracked pedestal under the statue of Apollo at his temple at Selinous. Rather than reproduce the method of Metagenes, however, Paconius proposed to modify it because of various design constraints. Since the ends of architraves and column shafts would be invisible in a completed temple, the pivot cups cut into their ends were of no matter. However, the ends of the pedestal would be in clear view, so it was not desirable to disfigure them merely for the sake of ease of transportation. Since the absence of pivot cups meant that a wooden hauling frame could not easily be employed, an alternate scheme had to be devised to pull the stone along. Wooden wheels could still be built around the pedestal, of course, and Paconius hit upon the idea of connecting two such wheels, each fifteen feet in diameter, with crossbars from wheel to wheel, thus enclosing the stone and forming a great wooden spool. As oxen pulled a rope wrapped around the spool, Paconius apparently imagined it would rotate along the ground and thus the important pedestal stone would be transported in a protective cage, as depicted in Figure 2.3. The rope might have been wrapped directly around the center of the stone, but that might not have occurred to Paconius. Or he may have felt that the rope would have abraded the pedestal's edges, or he may have perceived a mechanical benefit in having the rope wound around a wheel of larger diameter.

It would appear that Paconius's proposed method was a design development that evolved naturally from the successful schemes of Chersiphron and Metagenes. Metagenes' method of moving architraves would certainly have worked to move the pedestal also, but the resulting pivot holes would have marred the pedestal. Paconius's first objective then appears to have been to alter Metagenes' scheme to avoid cutting into the stone. Of course, the wooden wheels could have been extended outward from and around the ends of the pedestal to carry pivot cups, and this might appear to have been the most natural modification to make. However, this would necessarily have widened the assembly even more than the method of Metagenes. Since the stone was being moved to replace a cracked pedestal at the site of an already existing temple, it is likely that there would have been narrow streets between buildings or other restricted passageways along the route from the quarry to the temple. Such constraints might have led Paconius to seek ways to narrow rather than widen the device.

Even had Paconius been asked at the time about how he came to think of modifying Metagenes' idea of using the pedestal as an axle to one of constructing a protective spool around the pedestal, he might not have been able to articulate his reasoning. A conceptual design such as Paconius's takes place in a nonverbal mode of thought (cf. Ferguson, 1977), and thus neither words nor logic is necessarily available for rationalization. However, once the flash, inspiration, leap of imagination, or whatever it might be called, occurs, then the designer's verbal and logical faculties are certainly available for com-

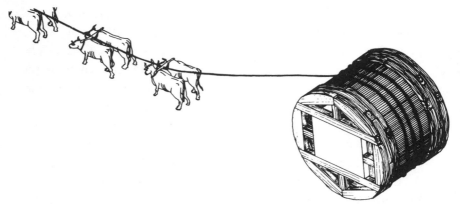

Figure 2.3. Paconius's scheme for transporting the pedestal for the statue of Apollo (Coulton, 1977; by permission of Cornell University Press).

municating with other engineers, clients, and workmen who might either criticize and modify or accept, finance, and execute the design. In the case of Paconius's scheme, there were plenty of reasons why it might have appeared to be a major breakthrough and thus one that should have been adopted without hesitation.

Not only did the scheme of Paconius keep the stone from being marred, it appeared to be a very economical method with several distinct advantages over that of Metagenes. For all its own brilliance and efficacy, after years of use the earlier method must have been recognized to have had some distinct disadvantages, as all familiar designs invariably do. First, the whole process of cutting holes in the ends of stones, fitting them with iron and lead inserts, constructing a mating pivot point in the wooden frame, and adjusting the whole device to turn freely and reliably, must have involved a much greater amount of skilled and coordinated effort and time than simply adding a few more timbers to connect the wheels. Second, the pivot assembly must have required regular maintenance and adjustment along the way from quarry to building site. Third, the unsupported stone axle between the large wooden wheels must have had to be pulled with some care so as not to be broken by a sudden jolt. (Some of the stones Metagenes transported over several miles were over twenty feet long and weighed as much as forty tons. There may well have been some mishaps at Ephesus, for when the stones were transported to Selinous, the wooden wheels were constructed not at the ends but at about a quarter of the way in from each end [Coulton, 1977], which would have reduced the maximum bending moment and stresses by half.)

The scheme of Paconius not only appeared to remove some of the objections that could be found to that of Metagenes, but also it must have seemed to have some distinct new advantages. In addition to the apparent savings in materials and labor required to assemble the device, it would appear that the spool technique had a mechanical advantage over the wheel and axle. If approximately the same size wheels were employed and the same number of oxen, the rope being pulled off the top of the spool would exert twice as great a turning moment on the load as the rope connected to the frame turning on the central pivots. Thus heavier loads could be moved with the same number of oxen, or fewer oxen could be employed to pull the same load. All in all, Paconius seemed to have happened upon a scheme that had every advantage over and no disadvantage with respect to the method of Metagenes.

The Scheme of Paconius: Its Realization and Failure Analysis

According to Vitruvius, Paconius modified the method of Metagenes "with confident pride," but the scheme did not work in practice the way it had promised to, and "Paconius got into such financial embarrassment that he became insolvent." We are not told by Vitruvius how or if the pedestal finally did get to its intended destination, but we are told the reasons why what appeared to be such a brilliant scheme failed. It was clear even two thousand years ago that knowing what went wrong with a design was perhaps even more important than what went right. Indeed, it was exactly the identification and

elimination of shortcomings in Metagenes' scheme that led to the
apparent improvements that Paconius conceived, but it was his neglect to foresee the shortcomings in his own scheme that led to its
ultimate failure to perform its function. As long as there are no
desires to improve designs, concentrating on success may be sufficient. But as soon as design changes are contemplated and made, it
is the correct understanding, perception, and anticipation of failure
that will save the designer from embarrassment, both financial and
professional.

What went wrong with the scheme that Paconius had devised?
According to Vitruvius, when the great timber spool had been constructed and a rope had been coiled around it, the oxen were yoked
up and driven to draw on the rope. However, "as it uncoiled, it did
indeed cause the wheels to turn, but it could not draw them in a
line straight along the road, but kept swerving out to one side."
Thus it became necessary to keep realigning the spool on the road.
(This had not been a problem in Metagenes' scheme because the
attachment of ropes to the ends of the frame provided a relatively
easy means of constant realignment.) Furthermore, since the rope
uncoiled from Paconius's spool at twice the rate that the spool advanced, the oxen would have had to be brought back and the rope
recoiled frequently. What appeared in principle to have been a brilliant and economical scheme turned out to be a disaster when it
came to practice. And neither Paconius nor any of his colleagues or
competitors seems to have foreseen the problem with the design or,
if they did not believe it would work, could not summon a logically
convincing argument to stop the process of design and implementation from advancing to embarrassment.

Although Vitruvius does not give many details about Paconius's
failed scheme, we may take the story as paradigmatic of the design
process and ask what might have been done to forestall the fundamentally erroneous conceptual design from getting as far as it did.
What if, for example, someone had asked Paconius to build a model

of his scheme to demonstrate how it would work? Clearly, as we can try for ourselves with a small spool of thread, it is difficult but not impossible to keep a spool rolling in a straight line without pulling the thread from side to side. Perhaps such a model demonstration could have been deflected by the confident Paconius, who might have argued that there would be such great weight, friction, and steadiness in the case of his large spool that the behavior of the model was irrelevant. The burden of proof would again be thrown back on the shoulders of the challengers. (Chapter 3 presents an example of how even models that work can be misleading.)

If a model had been employed, it would also have made evident that, as the spool rolled along the ground, the oxen would distance themselves in front of it at a rate equal to the progress of the spool. This might have raised the question of what might have happened if there were curves or turns in the road, for the rope would necessarily remain taut and obey no curbs. Perhaps Paconius or someone playing the role of error eliminator had thought of this logical objection to the scheme, and perhaps that is why a relatively short piece of rope may have been used to coil around the spool. But if that redesign response to the objection of a long rope had been tried on a model spool, then it might have made evident the fact that the spool would only advance a short distance before the oxen had to be stopped and brought back to the spool while the cord was recoiled around it. Since starting the heavy spool from rest (along with stopping it) might have been one of the more difficult parts of the operation, frequent restartings would have been clearly undesirable.

It would seem that Paconius neither used a representative physical model nor thought through much of a mental model for his scheme. Furthermore, it would seem that no one else involved in the moving of the great pedestal foresaw the problems with Paconius's method or, if they did, either they could not convince him or anyone else with a sufficiently articulate or logical argument that the thing would not work, or they decided to let the arrogant Paconius embarrass himself. Whatever actually happened in the case of Vitruvius's contemporary, however, there is clearly a great deal to be learned about design and error in the story of Paconius and the stone pedestal.

A Modern Example

Seemingly sound concepts for improving on familiar ways of doing things continue to hold surprises. One of the most promising new schemes for building high-rise apartment buildings evolved in Europe during the early 1960s (Great Britain Department of the Environment, 1973). It evolved naturally from conventional construction and comprised a system of assembling large factory-fabricated concrete components that minimized on-site construction work, an age-old source of inferior worksmanship and costly delays. Walls, floors, and stairways were all precast, transported to the construction site, and assembled in one-story-high units, all of which were load-bearing. By 1968 there were about 3,000 such apartment units in the British Commonwealth, including a twenty-four-story group of flats at Ronan Point, in East London. Early in the morning of May 16, 1968, a freak gas explosion blew out one of the corner walls of an eighteenth-floor flat, and a chain reaction ensued. The insufficiently supported rooms above fell in, and the impact caused the floors below also to collapse. One whole corner of the building was totally destroyed (see Fig. 2.4), three people were killed, and eleven others injured. The official report of the inquiry into the structural failure quoted a comprehensive building code that warned of "the absolute necessity of effectively joining the various components of the structure together in order to obviate any possible tendency for it to behave like a 'house of cards,'" but no code at the time specifically covered system buildings, as the prefabricated structures were called.

Conceptually, the system building was a brilliant alternative to costly on-site construction, and it could be argued that it allowed

Figure 2.4. The Ronan Point apartments collapse (Great Britain Department of the Environment, 1973).

for the achievement of better quality control in the individual components. However, as the gas explosion revealed, the system had a fundamental flaw in its design, and the loss of one load-bearing wall under the wrong conditions allowed one whole corner of the structure to collapse. The concept of system construction had many advantages over the more traditional schemes, but it clearly had the disadvantage of very little redundancy, for when one wall was blown out there was nothing left to support the walls above. The fundamental error of the concept was not revealed in the logic of the design process but in the chance events that led to a gas explosion. When an explosion blew out some upper walls in a more conventionally constructed apartment building in New York's Harlem in 1991, the overall building withstood the structural trauma, as it was expected to.

Not all errors in conceptual design lead to such dramatic collapses as that at Ronan Point. But many products do fail to reach their promised potential because they seem not to have been sufficiently criticized or successfully scrutinized before being launched as full-scale structures, full-power machines, or full-blown products. It may be argued that virtually all design errors fall into the category epitomized by the story of Paconius attempting to move the pedestal, in the sense that any error in design as opposed to one in manufacture, maintenance, or operation is based on a flawed concept that before failure occurred was not detected to be crucial to the design. If so, then it would appear to be incontrovertible that understanding and focusing on the design errors of the past, no matter how far distant and no matter how different in application, can be beneficial for reducing human error in today's designs. The more case histories of failure that designers and design critics can summon up in discussions about whether a new concept is sound, the more sound and the fewer flawed new concepts there are likely to be.

3
Vitruvius's Auger and Galileo's Bones
Paradigms of Limits to Size in Design

Perhaps no principle in design is so well known and yet so frequently forgotten as the effect of size or scale on performance. Although some of civilization's most proud moments have been immortalized in colossal monuments to the triumph of human design ingenuity over the great forces of nature, some of history's most embarrassing moments have come in the dramatic failures of some of the largest machines, structures, and systems ever attempted. Nowhere does the line between success and failure seem so thin as when the very greatest of our human efforts straddle it.

Some of the earliest recorded thought about design shows an awareness of a scale effect in natural and fabricated things. Among the minor works of Aristotle is a collection of questions and answers known as *Mechanical Problems,* in which queries about the physical world are posed and solutions offered in the context of fourth-century B.C.E. knowledge of physics, mathematics, and engineering. Although there is some doubt that *Mechanical Problems* is actually the work of Aristotle himself, rather than of his Peripatetic School, certainly the work fairly represents contemporary thinking and shows that engineers of twenty-four centuries ago wrestled with problems that can still be troublesome today, as indicated by recurring design errors and structural failures. Among the thirty-five mechanical problems posed in Aristotle's time were "Why [are] larger balances . . . more accurate than smaller ones?" and "Why are pieces of timber weaker the longer they are, and why do they bend more easily when raised?" The answers given may not be sufficiently rigorous by today's standards, but the mere fact that the questions were asked and the manner in which they were answered suggest that they arose out of uncertainties and difficulties encountered by ancient engineers in scaling up successful designs. Three centuries after Aristotle's problems, Vitruvius's compendium of the state of the state of the art of ancient architecture and engineering also contained clear warnings about the limits of magnification of machines and structures.

Although our knowledge of physics, mathematics, and the engineering sciences has obviously advanced since the time of Aristotle and Vitruvius, there appear to be essential features of engineering design that remain fundamentally unchanged. These include the nature of the conceptual design process itself, in which concepts arise as alogical acts of creativity rather than as totally logical acts of deduction. Only after they are conceived can design ideas be subjected to the rational analysis that is supposed to filter the bad from the good ideas and modify promising but imperfect schemes into workable and reliable artifacts. But, as everyone knows, the filtering and modifying aspects of the design process are far from perfect, and engineers have always been and continue to be embarrassed by design errors that manifest themselves in ways ranging from inconvenience to catastrophic failure.

Design errors attributable to overlooking scale effects in nature and artifacts have been especially persistent throughout history, and they continue to be so even in our age of high technology and computers. Although it had been known for more than two thousand years that pieces of timber grow weaker as they grow longer, Renaissance shipbuilders found inexplicable the fact that their large timber ships were breaking under their own weight. Galileo (1638) prefaced his seminal study of strength of materials by reciting the breakup of ships and other recurring failures of Renaissance engineering attributable to size, a problem that he noted nature had well under control. The story of Gothic cathedrals and the history of bridge building may be related with an underlying theme of how limits to size were defined by recurring structural failures. And even in the late twentieth century, failures of heavy steel sections (see, e.g., Fisher, 1984) and large missiles (e.g., Rosenthal, 1989) have been attributed to design errors in overlooking or underestimating the effects of size in scaling up successful designs.

There is clearly no guarantee of success in designing new things on the basis of past successes alone, and this is why artificial intelligence, expert systems, and other computer-based design aids whose logic follows examples of success can only have limited application. It is imperative in the design process to have a full and complete understanding of how failure is being obviated in order to achieve success. Without fully appreciating how close to failing a new design is, its own designer may not fully understand how and why a design works. A new design may prove to be successful because it has a sufficiently large factor of safety (which, of course, has often rightly been called a "factor of ignorance"), but a design's true factor of safety can never be known if the ultimate failure mode is unknown. Thus the design that succeeds (i.e., does *not* fail) can actually provide less reliable information about how or how not to extrapolate from that design than one that fails. It is this observation that has long motivated reflective designers to study failures even more assiduously than successes.

If the effect of scale has been an especially troublesome one throughout the history of design, then it follows that much may be learned from failures attributable to it. Two classical discussions, one by Vitruvius and one by Galileo, of failures associated with increasing size are explicated here as paradigms for the scale effect.

Vitruvius on Scale

In the closing book of his *Ten Books on Architecture,* Vitruvius introduces the idea of a scale effect by telling the story of an engineer named Callias who came to Rhodes and gave a public lecture in which he demonstrated a model of a revolving crane that was designed to sit on the city's wall and capture any enemy siege machine that might approach. The Rhodians were so impressed that they gave Callias the annual retainer that had formerly gone to one Diognetus, whose schemes for the defense of the city followed more conventional practices.

In the meantime, war was made on Rhodes, and a huge siege machine, known as a helepolis and weighing over 150 tons, was deployed by the enemy. Callias was asked to construct a full-scale version of his model so that it might capture the helepolis and bring it within the wall where it could no longer threaten the Rhodians, but he shocked his supporters by saying that he could not do so because it was impossible. Vitruvius explains why Callias could not scale up his model crane:

> For not all things are practicable on identical principles, but there are some things which, when enlarged in imitation of small models, are effective, others cannot have models, but are constructed independently of them, while there are some which appear feasible in models, but when they have begun to increase in size are impracticable, as we can observe in the following instance. A half inch, inch, or inch and a half hole is bored with an auger, but if we should wish, in the same manner, to bore a hole a quarter of a foot in breath, it is impracticable, while one of half a foot or more seems not even conceivable.

We may no longer use augers to bore holes, but we are familiar with the analogous problem of driving a screw. The larger the screw, the larger the screwdriver we use, of course, but for screws with heads approaching even a half-inch it becomes extremely difficult, if not impossible, for the ordinary person to drive the screw in hard

wood that is not predrilled. Larger screws tend to have square or
hexagonal bolt heads, so that we may gain the much larger mechanical advantage of a wrench.

Vitruvius continues the story of Callias by explaining how the
Rhodians had inflicted injury and insult on Diognetus because they
were deceived by false reasoning – specifically, believing that what
worked on the scale of a model could also work on any larger scale.
With their city threatened by the mammoth siege machine, the Rhodians pleaded with Diognetus to help them. At first he refused, but
later agreed, and Vitruvius goes on to tell the story of Diognetus's
triumph:

> He pierced the wall in the place where the machine was going to ap-
> proach it, and ordered all to bring forth from both public and private
> sources all the water, excrement, and filth, and to pour it in front of the
> wall through pipes projecting through this opening. After a great amount
> of water, filth, and excrement had been poured out during the night, on
> the next day the helepolis moving up, before it could reach the wall, came
> to a stop in the swamp made by the moisture, and it could not be moved
> forwards, nor later even backwards.

The enemy retreated, naturally, and Diognetus was honored. He
brought the helepolis into the city and set it up in a public place as
a reminder to the people. From this story Vitruvius draws a moral
about how to construct works of defense, but that is clearly not the
value of the story for the nuclear age. Rather than teaching us how
to design deterrents, the story has its lasting value in indicating what
can be overlooked in design generally. Clearly the helepolis might
not have worked even in a heavy rain, and so its own limitations
were inherent in its great bulk, even though a model of the machine
might have worked perfectly on a clear day.

The story of Diognetus and the helepolis has its greatest value
for modern design in pointing out the limitations of size in a memorable way. The very details of the story, the names of the designers
and the commonness of the materials used to bog down the helepolis,

serve to make it real and thereby more effective. This paradigm is not an anonymous and dry rule such as "models cannot be scaled up indefinitely," but rather a human drama full of sights and smells. Such associations with design·problems should not be seen as distracting so much as making the lessons more effective. If paradigms such as this can become more regularly a part of the lore and language of engineering, so that the mere mention of Callias or Diognetus, for example, brings to mind the cautions that must go with scaling up designs, then perhaps the ageless caveats about scale will be more heeded and large-scale errors in design can be averted. If vivid case studies of failure attributable to scale effects do nothing else than cause designers to treat with more suspicion their scaled-up versions of successful designs, there is reason to believe that the mistakes of the past will not be thoughtlessly repeated.

Galileo and the Scale Effect

When Galileo's *Dialogues Concerning Two New Sciences* was published in 1638, it represented its septuagenarian author's thinking about some problems that he had considered on and off for decades. The bulk of the book takes the form of four days of discussion among the interlocutors Salviati, Sagredo, and Simplicio, who, respectively, represent the mature Galileo, a younger Galileo, and their naive foil. The book is referred to by historians of science as the *Discoursi,* to distinguish it from Galileo's earlier work, *Dialogue Concerning the Two Chief World Systems, Ptolemaic and Copernican,* and is generally known as a book on motion. Here, however, where there is little chance of confusing the two works anyway, Galileo's mature work will be referred to as *Two New Sciences,* and all quotations from it will be from the Crew and de Salvio translation of the original Italian, whose title page is reproduced as Figure 3.1.

The first two days of dialogues in *Two New Sciences* constitute Galileo's seminal work on the engineering science of strength of

DISCORSI
E
DIMOSTRAZIONI
MATEMATICHE,
intorno à due nuoue ſcienze
Attenenti alla
MECANICA & i MOVIMENTI LOCALI;
del Signor
GALILEO GALILEI LINCEO,
Filoſofo e Matematico primario del Sereniſſimo
Grand Duca di Toſcana.

Con vna Appendice del centro di granità d'alcuni Solidi.

IN LEIDA,
Appreſſo gli Elſevirii. M. D. C. XXXVIII.

Figure 3.1. Title page of Galileo's seminal work on strength of materials and dynamics (Galileo, 1638).

materials. But rather than describe the great structures that might be built with the techniques he is about to reveal, Galileo opens the first day of dialogues with descriptions of some embarrassing structural failures and with a clear demonstration that the Renaissance state of the art, which employed pure geometrical reasoning to scale up artifacts, was insufficient to design successfully and with full understanding of their behavior very large engineering structures.

It is Galileo echoing Vitruvius when Sagredo remarks that "one cannot argue from the small to the large, because many devices which succeed on a small scale do not work on a large scale," but he confesses that he cannot explain why from geometrical principles alone. Salviati acknowledges that a scale effect was commonly attributed to "imperfections and variations of the material," but dismisses this explanation and asserts that "imperfections in the material, even those which are great enough to invalidate the clearest mathematical proof, are not sufficient to explain the deviations observed between machines in the concrete and in the abstract." And he makes Sagredo's brain reel by stating that "for every machine and structure, whether artificial or natural, there is set a necessary limit beyond which neither art nor nature can pass [when] the material is the same and the proportion is preserved."

Salviati presents the example of a cantilever beam made of a given kind of wood and declares that for a given cross section there is a unique length for which the beam will just support itself: if it were any longer it would break, but if it were any shorter it could support a load in addition to its own weight. Furthermore, the scale effect manifests itself in the fact that while "a piece of scantling will carry the weight of ten similar to itself, a beam [of the same wood but much longer and yet] having the same proportions will not be able to support ten similar beams." Nature understands this, we are reminded with numerous examples, and thus with regard to artifacts the belief that "the very large and the small are equally feasible and lasting is a manifest error." But while a scale effect was known in Galileo's time, the principles behind it were not, as demonstrated by the story Salviati tells of the marble column (related below in Chapter 4) that broke after precautions were taken to obviate exactly that.

Questions of size are resolved by understanding how things break, and finding the critical length of a cantilever beam becomes the principal object of Galileo's analysis, but it appears only after much discussion of the resistance of solid bodies to fracture and the causes of their cohesion. By attacking the fundamental problem of the cantilever beam, albeit with some erroneous assumptions about its failure mode (as discussed in Chapter 5 below), Galileo is able to articulate and formulate rational strength-of-materials solutions to problems of design and optimization, such as what size of timber should be used to support a given weight at a given distance from a wall and what profile a cantilever beam should take in order to be critically stressed at each point along its length. And he is able to answer a question posed in the *Mechanical Problems* of the Peripatetics: "Why is a piece of wood of equal size more easily broken over the knee, if one holds it at equal distance far away from the knee to break it, than if one holds it by the knee and quite close to it?"

Having devised a rational approach to determining the strength of materials, Galileo was able to give answers to many of the questions that had bothered thinkers long before and long since Aristotle. Because Galileo had determined principles by which he could calculate failure in the materials and structures of natural and artificial things, he was able to explain mysteries and paradoxes of long standing and was able to state his achievement:

From what has already been demonstrated, you can plainly see the impossibility of increasing the size of structures to vast dimensions either in art or in nature; likewise the impossibility of building ships, palaces, or temples of enormous size in such a way that their oars, yards, beams, ironbolts, and, in short, all their other parts will hold together; nor can nature produce trees of extraordinary size because the branches would break down under their own weight; so also it would be impossible to build up the bony structures of men, horses, or other animals so as to hold together and perform their normal functions if these animals were to be increased enormously in height; for this increase in height can be accomplished only by employing a material which is harder and stronger than usual, or by enlarging the size of the bones, thus changing their shape until the form and appearance of the animals suggest a monstrosity.

The phenomenon of the scale effect is perhaps best summarized in Galileo's illustration of "a bone whose natural length has been increased three times and whose thickness has been multiplied until, for a correspondingly large animal, it would perform the same func-

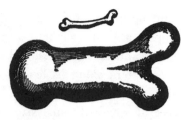

Figure 3.2. Galileo's illustration of nature's appreciation of a scale effect (Galileo, 1638).

tion which the small bone performs for its small animal." The famous illustration is shown here as Figure 3.2, and it may serve as a reminder that the effect of size, like all effects in engineering, can be understood fully only when one fully understands, following Galileo, how a bone can break.

The Persistence of Scale Effects in Shipbuilding

In shipbuilding especially, problems associated with scale effects have continued to be uncannily persistent throughout history, suggesting an ignorance or dismissal of lessons learned. At the time of Galileo the problem seems to have been especially acute and visible, so much so that it is employed on the first page of his introduction to the study of strength of materials to demonstrate the inadequacy of geometry alone to predict the success of scaling up a proven design. Contemporary with Galileo's experience in Italy, a newly built Swedish ship of unprecedented size, the *Vasa*, was shown to be top-heavy in the water when its stability was tested by having a group of sailors run back and forth across its deck. (The tale of this "tender ship" has been taken by Squires, 1986, as paradigmatic not so much of a structural scale effect as of problems associated with large technological projects.)

Problems with large ships arose also in the mid-nineteenth century, when iron was replacing timber as a shipbuilding material, and the British engineer Isambard Kingdom Brunel experienced considerable difficulty in launching the monumental *Great Eastern* in 1857 (see, e.g., Rolt, 1970; Petroski, 1992a; but cf. Emmerson, 1977; Vaughan, 1991). Fairbairn (1865) and Russell (1864) wrote extensively on ships as beam structures, and Russell especially was quite explicit about seeing no limits to their size or speed:

I am one of those who believe, that neither the speed, the economy, nor the security and comfort of ocean navagation, have reached their limit. In my youth, 10 knots an hour were reckoned the *ne plus ultra* of possible steam navigation, – seldom attained in smooth water, and then by rare and happy accident. Her Majesty's first steam yacht was thought to do great things, when she did that. Then came 12 knots an hour; next, 14 knots: 16 and 18 are now attainable with perfect certainty; and I venture to say, that 20 knots across the ocean is a speed, which can be readily attained, whenever it is required and believed in. Indeed, with iron ships, I fail to see the limit of safe speed in anything, except the serious matter of cost. Whatever speed trade will remunerate, or society choose to pay for, they can henceforth attain on the sea with precision and security.

The optimism and enthusiasm of Russell were subsequently dramatized near the turn of the century in a world's fair exhibit on Cunard steamers that, although not "thoroughly worked out," caught the attention of the historian Henry Adams. Writing in his autobiography in the third person, he reflected philosophically about the limits to the size of great ships, and he complained that "a student hungry for results found himself obliged to waste a pencil and several sheets of paper trying to calculate exactly when, according to the given increase of power, tonnage, and speed, the growth of the ocean steamer would reach its limits. His figures brought him, he thought, to the year 1927" (Adams, 1918). Soon, of course, the likes of the *Titanic* were confidently being launched – and sunk – amidst great confusion. Seventy-five years after its inaugural and final Atlantic crossing, the *Titanic*'s true failure mode was still open to question by some when no large gash was found on the newly discovered wreck (Sullivan, 1986). It appeared that the mere size of the great ship may have had as much to do with its failure as the sharpness of an iceberg. Nearly contemporary with the *Titanic*, rigid airships were also reaching their limits to growth, as the fate of the R100 and R101 dirigibles so convincingly demonstrated (cf. Shute, 1954; Squires, 1986).

When welding replaced riveting in the fabrication of large steel oceangoing vessels and Liberty ships were produced in record time to transport supplies during World War II, new effects of size in material behavior contributed to the surprise appearance of brittle fracture (cf. Great Britain Navy Department, 1970; Richards, 1971; and see Fig. 4.5). Even today, as supertankers ply the seas with millions of barrels of potentially polluting cargo, the mysterious breakup of some ships is of ongoing concern and debate (Anonymous, 1989; cf. Sullivan, 1986). Gordon (1988) has been especially articulate and blunt about this matter, pointing out that a certain absolute size of crack in a steel hull may be considered relatively large or small, depending on the size of the ship it is found in (cf. Fisher, 1984, for a similar discussion for steel bridges). However, a relatively small crack may be of critical length:

Because small and medium-sized metal structures have proved to be safe in service, engineers and naval architects have scaled them up, assuming that if the stresses remained the same the large structure would be as safe as the smaller one. But the critical crack length in the mild steel plates used in engineering is between 1 and 2 meters [regardless of the size of the structure in which it occurs].

In other words, as the size of ships increases, relatively smaller cracks may be considered the limiting factors in whether the vessels can hold together in rough or sometimes even calm seas. Given the special difficulties that come with increasing size, it would seem that a familiarity with the history of shipbuilding and its most famous failures would be as important to the modern designer as mastery of the rules of naval architecture.

Figure 3.3. The Menai Strait (1826), Brooklyn (1883), and George Washington (1931) suspension bridges drawn to the same scale (*Civil Engineering*, August 1933; by permission of the American Society of Civil Engineers).

Scale Effects in Bridges

There are classic examples of good practice in the design and erection of bridges and bridgelike structures of enormous ambition and scale, most notably the Britannia Bridge across the Menai Strait in northwest Wales (discussed in Chapter 7) and the contemporaneous Crystal Palace erected for the first world's fair, held in London's Hyde Park in 1851 (see, e.g., Petroski, 1985). But there have also been some colossal errors. The history of suspension bridges provides an especially sad chronicle of how the mistakes of the past can be repeated in a state-of-the-art environment that appears to have more confidence in its analytical sophistication than fear of the effects of scale as bridge spans have increased in bold leaps rather than incremental steps (see Fig. 3.3). Such overconfidence led, of course, to the classic failure in 1940 of the Tacoma Narrows Bridge, along with the excessive flexibility of several of its contemporaries (cf. Petroski, 1987a, and see Chapter 9 below). The phenomenon was repeated also in the history of large cantilever bridges (see, e.g., Sibly, 1977), culminating in the embarrassment of the (first) Quebec Bridge across the St. Lawrence River, which collapsed during construction in 1907 (see Figs. 3.4 and 3.5). Although the span of the Quebec Bridge was essentially the same as that of the colossal Firth of Forth Bridge (see Fig. 6.7), which was of a daring span when completed in Scotland in 1890, some of the Quebec's compression

Figure 3.4. The Quebec Bridge under construction in 1907 (Canada, 1919).

members were so slender that they could fit inside the corresponding members of the Forth Bridge, as shown in Figure 3.6. The belief that the stocky Forth Bridge was very much overdesigned generated a confidence that paring down the members of the Quebec span carried little risk. In effect, the bones of the bridge were too slender to carry its own weight (cf. Fig. 3.2). The structure was subsequently redesigned with heavier members arranged in a modified pattern, and the (second) Quebec Bridge stands today (Fig. 3.7) as a symbol to Canadians of perseverence in the face of adversity (cf. Canada, 1919).

It is not necessary to look only to the extreme cases of monumental bridges in order to find misunderstandings of scale effects. Steel-girder bridges of more pedestrian design have experienced cracking and fatigue problems in recent years, and many have been due to the excessive scaling up of successful designs by merely using

Figure 3.5. The collapsed Quebec Bridge, 1907 (Canada, 1919).

heavier steel sections (cf. Fisher, 1984). According to John Fisher, the fatigue and fracture expert and failure analyst who was named Man of the Year in 1987 by *Engineering News Record* (Tuchman, 1987), "Scale effects are one of the major engineering problems we face. Designers are prone to extrapolate beyond the existing knowledge base." Furthermore, he believes that an alarming number of substantial structural failures in this age of high technology are a result of the lack of research and development in such areas as new fabrication techniques.

Another instance involved a massive steel truss in the Orlando Civic Center. It cracked, according to Fisher, because designers assumed that the full-size truss would behave as well as a small-scale laboratory model. He concluded: "That came about from a complete lack of knowledge about scale effects. When they made a steel section

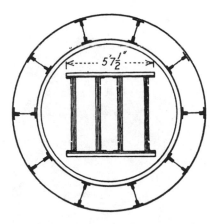

Figure 3.6. Bottom chord members of the Firth of Forth Bridge (circular section) and the Quebec Bridge (rectangular section), drawn to the same scale (*Engineering News*, September 26, 1907).

bigger, it had characteristics that were completely different from when it was smaller" (Pechter, 1989).

The design and erection of large and tall structures have always derived much inspiration and experience from bridge building, and significant high-rise structures ranging from the Eiffel Tower (erected for the international exposition held in Paris in 1889) to such contemporary Chicago skyscrapers as the John Hancock Building (completed in 1970) and the Sears Tower (1974) owe much of their success to thoughtful considerations of scale in the use of ingenious and novel structural forms. However, as in the history of bridges, there are examples of tall buildings that have been designed at the limits of size but have demonstrated what would appear to be an insufficient consideration of scale effects. Although there have not been catastrophic collapses of very tall buildings, there have been notable performance problems, as epitomized by the early behavior of the John Hancock Tower in Boston, whose unanticipated torsional flexibility led to problems with its windows, whose large panes of glass were prone to popping out and falling to the street below (see, e.g., Marlin, 1977; Campbell, 1988).

Figure 3.7. The (second) Quebec Bridge, as completed in 1919 (Canada, 1919).

Conclusion

The problem of scale effects has long proven to be an especially persistent one, and understanding size limits is central to the advancement of the state of the art of design. New methods of analysis and new theories have long been offered as antidotes to the ignorance that has led to failures due to scale effects, but applying new theories without understanding the significance of old experience can be disastrous. In 1856, the Scottish engineer and educator W. J. M. Rankine wrote about the necessity of a "harmony of theory and practice," the knowledge of which

qualifies the student to plan a structure or machine for a given purpose, without the necessity of copying some existing example, and to adapt his designs to situations to which no existing example affords a parallel. It enables him to compute the theoretical limit of the strength or stability of a structure, or the efficiency of a machine of a particular kind, – to ascertain how far an actual structure or machine fails to attain that limit, – to discover the causes of such shortcomings, – and to devise improvements for obviating such causes; and it enables him to judge how far an established practical rule is founded on reason, how far on mere custom, and how far on error.

Scale effects that concerned the ancients no less than they seem to have Rankine in the nineteenth century are still the cause of failures today, and this appears to be in large part because designers who might be relying too much on mathematical and computer models are lacking in a clear and full understanding of the physical phenomenon of a scale effect or do not treat it with due respect. In short, they seem to embrace theory without fully harmonizing it with practice.

4
Galileo and the Marble Column
A Paradigm of a Design Change for the Worse

Galileo begins the first day of *Dialogues Concerning Two New Sciences* by relating several stories of contemporary structural failures that were inexplicable within the state of the art of Renaissance engineering. As indicated in the last chapter, several of the stories have to do with the problem of proportionately similar structures behaving differently depending upon whether they are large or small, hence manifesting a scale effect. According to Galileo,

Thus, for example, a small obelisk or column or other solid figure can certainly be laid down or set up without danger of breaking, while the very large ones will go to pieces under the slightest provocation, and that purely on account of their own weight.

Such generalizations could, of course, be reached by experience alone, and there were no doubt even rules of thumb as to how large an obelisk or column had to be to require special care in its storage or erection. Empirical evidence of failure and rules of thumb were not sufficient to explain why a failure would occur, of course, but they could certainly help avoid failures in situations recognized to be precarious. Indeed, with little more than an empirical understanding of what worked and what could go wrong when the untried was

47

Figure 4.1. Proposals to move the Vatican obelisk in 1586, as illustrated in Fontana's 1590 book *Trasportatione dell' Obelisco* (Dibner, 1950).

Figure 4.2 Marble column in storage: *top*, with modified support; *bottom*, as originally supported (after Fung, 1977).

tried, Renaissance designers might have been particularly cautious in their attempts to extrapolate to larger structures and ever attentive to situations that might lead to failure. When the Vatican obelisk was to be moved to the front of St. Peter's in Rome, for example, there were numerous proposals as to how it might be done without mishap (see Fig. 4.1; cf. Dibner, 1950).

Among the stories that Galileo's alter ego, Salviati, relates to his interlocutors Sagredo and Simplicio early in the first day of *Two New Sciences* is what he promises them to be "a circumstance which is worthy of your attention as indeed are all events which happen contrary to expectation, especially when a precautionary measure turns out to be a cause of disaster":

A large marble column was laid out so that its two ends rested each upon a piece of beam; a little later it occurred to a mechanic that, in order to be doubly sure of its not breaking in the middle by its own weight, it would be wise to lay a third support midway [as in Fig. 4.2]; this seemed to all an excellent idea. . . .

And this is an excellent paradigm of the design process. The column resting on the pieces of beam (or rocks, as shown in the modern interpretation of Fig. 4.2) is the given design, and the observant mechanic is the conscientious checker, who might even be the original designer himself. As a good checker should, he is thinking of ways in which the column might fail to function as intended, in this case simply to rest in storage without breaking. However, knowing that large columns (like obelisks) can break under their own weight, the checker gets the idea of reducing the probability of failure by increasing the safety of the simple structure, perhaps reasoning that adding a third support will reduce the load that the column's material itself will have to bear across the span between its ends. Good practice dictates that the checker discuss his proposed design change with his peers and superiors to be sure that he is not overlooking something in what seems to be a clear improvement, and good practice also requires that he have the change approved by the engineer-in-charge. He presents his proposed change to them, and they all agree it is an excellent idea. Thus the third support is added. And everyone is relieved that the designed storage system is more reliable, that the column is now more safely stored than it was in the original configuration.

(The reader interested in second-guessing the Renaissance decision makers, having the advantage of knowing that their design change proved not to be efficacious, might wish to think about the problem of the marble column at this point. How was the piece of marble to break? What was it about the altered means of support that was to prove disastrous to the column? What changes might have been made to avoid disaster? And, finally, would the reader have been as likely to have objected to the change that was proposed or have called for any further changes without the special knowledge that the very thing that steps were taken to avoid, namely, the failure of the column, was in fact going to occur?)

A Paradigm of Error

Since they thought they had solved the problem to their satisfaction, the Renaissance mechanic and his co-workers no doubt gave little more thought to the marble column, other than perhaps admiring their own sagacity each time they passed the column resting unremarkably in the yard. In fact, because they were so proud of their improvement in the design of its added support, we can imagine that it was probably that single detail that would most have caught their eye as they walked past on their way to other more pressing design or construction problems. What occupied their thoughts might have been the much larger column being erected in the cathedral under construction or the great obelisk being moved from one piazza to another. Indeed, they no doubt soon came to ignore and forget altogether the stored column.

But Salviati was not finished with the story of the marble column. Everyone's relief after the addition of the third support was soon to be everyone's grief, for the column was to break. Its third support had certainly "seemed to all an excellent idea . . . but the sequel showed that it was quite the opposite, for not many months passed before the column was found cracked and broken exactly above the new middle support."

Simplicio, who in *Two New Sciences* generally speaks for the old tradition, acknowledges that the failure of the column was a "very remarkable and thoroughly unexpected accident, especially if caused by placing that new support in the middle." Salviati agrees that this was indeed the cause, but he tells Simplicio and Sagredo that "the moment the cause is known our surprise vanishes." Salviati then goes on to explain how the failure occurred, and his description of the sequence of events remains to this day a model of failure analysis:

[W]hen the two pieces of the column were placed on level ground it was observed that one of the end beams [supports] had, after a long while, become decayed and sunken, but that the middle one remained hard and strong, thus causing one half of the column to project in the air without any support. Under these circumstances the body therefore behaved differently from what it would have done if supported only upon the first beams; because no matter how much they might have sunken the column would have gone with them.

In other words, the very thing that first worried the conscientious mechanic, that the stored column could not support its own weight, proved to be a legitimate concern. The mechanic foresaw that the column could break by a crack opening up at the bottom and the two halves falling inward between the extreme supports, but he did not foresee that changing the means of support affected the whole structural system and introduced the possibility of a new mode of failure, one in which one half of the column projected unsupported like a cantilever. The original support system might indeed have been marginally safe, but the natural placement of the supports somewhat in from the ends of the column made its weight a less crucial loading condition than the one that existed over a carefully designed and centered third support when the original supports settled. If, for example, the original supports were at the extreme eighth points of the column, the maximum bending stress in the revised design could have been as much as twice what it was in the unsupported middle of the original design. Galileo's own illustration of the two failure modes, which he did not present until after he had analyzed them fully to his satisfaction in the second day of *Two New Sciences,* is shown in Figure 4.3.

The Design Problem and Failure Avoidance

The error related by Galileo is a common one in design: starting to analyze a problem in the middle and forgetting to go back to the beginning. Every solution of every design problem begins, no matter how tacitly, with a conception of how to obviate failure in all its possible and potential manifestations. Although what must be understood before all else in the formulation of the original problem is what is to be achieved and what is to be avoided, both desired ends and undesired consequences are best dealt with in terms of failure avoidance.

In the case of the marble column, the positive objectives may have

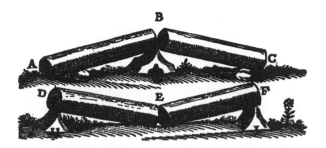

Figure 4.3. Galileo's illustration of two failure modes (Galileo, 1638).

included keeping the column clean and in a position from which it could be easily removed when needed, while being kept safely intact in the meantime. These positive objectives are more operationally described in terms of their negatives: the design objectives for a column storage system were that the column must not rest on the damp ground and thus possibly become discolored, it must not be difficult to move when needed, and it must not be in danger of being broken. Experience, and virtually all of it would have been bad experience, would have included various negative observations. A marble column lying directly on the ground could become soiled or moldy, and it would be hard to get hands or levers underneath it to pick it up without doing damage. A column stored in a more upright position, such as inclined against a wall, would be likely to crack like an unsupported obelisk or to fall over and break. Given some bad experiences, it is not hard to imagine how the concept of storing a column horizontally on two supports could arise: it kept the column off the damp ground and also made it easier to get hands and lifting devices under it when needed. Furthermore, a horizontal column was not likely to topple over and could be fluted or otherwise worked in a position convenient for the stone cutter.

It is likely that columns had been stored and worked in that position for ages before the Renaissance. However, negative experience may have been accumulating, including the breakage of increasingly large columns, some of them even just giving way in the middle while in storage, or perhaps failing when a group of workmen sat on them to eat lunch or when a particularly hard blow from a mallet fell on a flawed piece of marble. In such a climate of increasing failures, it is not surprising that a mechanic might have gotten the idea of improving the existing two-support system to reduce the possibility of the failure of a column for which he had some responsibility.

As with so many modifications of design concepts, however, the change was apparently conceived and made without thinking much beyond how it improved only the one aspect of the original design that was observed to be lacking. Since the column on two supports was imagined to be in danger of breaking under its own weight, the idea of transferring some of the weight to a support where the break might occur would seem logical. However, good design does not proceed on such narrow considerations, and it is the responsibility of the redesigner to consider how the alteration interacts with the rest of the existing system and how the alteration satisfies the original objectives of the design problem. In the case of the marble column, the idea of adding a third support should have been followed by deliberate considerations of whether the newly supported column could get more dirty, could be harder to pick up, or could break in a different way.

The Paradigm and Other Designs

Galileo prefaces his seminal inquiry into the strength of materials not only by telling the story of the marble column but also, as indicated in Chapter 3, by referring to the breakup of large wooden ships while they were being launched. Since Renaissance shipbuilders knew that this was occurring, they took extra precautions with large new ships designed on purely geometrical principles but had no theory of strength of materials to explain why a scaled-up version of a successful and hardy small boat should behave so differently. Galileo further recognized the fundamental similarity among the problems of wooden ships and marble columns and the cantilever beam (which will be discussed in Chapter 5) that he made the principal focus of his analysis in the second day of *Two New Sciences*. It is an invaluable aid to designers to remember with Galileo that seemingly different objects can behave in astonishingly similar ways. And the ability to draw broad analogies can be an invaluable aid to the anticipation of failure in novel designs.

In the nineteenth century large ships came to be made of iron, and in his 1865 *Treatise on Iron Shipbuilding* William Fairbairn illustrated the extreme ways in which a ship could be supported on waves, here reproduced as Figure 4.4. The observation that under certain sea conditions a great part of the weight of a ship could essentially be resting alternately on two waves at its ends and on a single wave under its midsection can suggest, with not too much of a leap of the imagination, Galileo's columns (as shown in Fig. 4.3). Analogous support conditions might also be encountered when a ship was stranded on rocks or sandbars or when it was being launched. Even without an analytical theory of structures by which one could calculate the intensity of stresses on a ship undergoing such treatment, the ship designer should have been able to see the danger of a ship breaking because of a weakness of the deck just as easily as because of a weakness in the bottom of the hull.

Because of his experience with the Britannia and Conway tubular bridges (see Chapter 7), and because he saw ships as large floating tubular beams, Fairbairn knew that their hulls should be as strong at the top as at the bottom, and not only in tension but also in compression lest crushing (or buckling) occur. He criticized contemporary practice, which was generally dictated by Lloyd's and other insurance underwriters, and recommended changes in the prevailing rules and regulations of ship design. In particular, Fairbairn proposed using the length as well as the capacity of the ship to determine the amount of material used in the hull, making it stronger at midships compared to the bow and stern, and putting equal material in the deck and hull.

Such design insights clearly led to structural improvements, but

Figure 4.4. William Fairbairn's illustration of ship loadings: *top*, supported on two wave crests; *bottom*, supported on a single wave crest (Fairbairn, 1865).

as the art of shipbuilding evolved over the next century, with new materials and techniques being introduced, new means of triggering old failure modes were also introduced. In particular, wrought iron gave way to steel and riveted to welded joints during World War II, and in the early 1940s many of the so-called Liberty ships suffered a fate illustrated in Figure 4.5. The resemblance to Galileo's broken column (Fig. 4.3) is striking. Structural mechanics had evolved by this time to enable the calculation of the stresses and strains capable of breaking a ship in two, but it was more subtle design changes that caused such catastrophic brittle fractures. The riveted joints of overlapping plates had served as effective arresters of any cracks that might begin to grow in the hull, and which could be repaired in due course. Welded construction had not only removed such obstacles from the paths of any cracks that might develop, but also the very

Figure 4.5. A failed Liberty ship, c. 1940 (Richards, 1971; by permission of The Welding Institute).

process of welding embrittled the adjacent steel and made it behave much like a ship of glass under circumstances such as led to the fracture in Figure 4.5.

For all our advances in analysis, the fundamental sequence of design questions is still ultimately (cf. Petroski, 1985): (1) How can failure occur? (2) What design feature can obviate that failure mode without introducing another? Paradigms like the story of Galileo's column can help keep these questions in the forefront, even in cases less obviously analogous than ship design.

Any design change, whether in geometry or material or process, can introduce new failure modes or bring into play latent failure modes. Thus it follows that any design change, no matter how seemingly benign or beneficial, must be analyzed with the objectives of the original design in mind. Although a structure designed the old way may be perfectly safe, an "improved" or enlarged design could hold very unpleasant surprises. This is not to say that old ways should never be changed, for that too would be irresponsible engineering. Indeed, one critic (Anonymous, 1989) of current shipbuilding practices has noted that in the decade of the 1980s about 160 large ships were lost at.sea, including some because of structural failure, and those failures may have been due to a lack of imagination or foresight in design. According to a report on these concerns, Richard Bishop, a naval architect himself, believed that rather than "sit down and design a ship and then test their ideas, [naval architects] can go for rules produced by Lloyd's which they know from past experience work." Unfortunately, when the designs of larger ships are based on rules extrapolated from experience with smaller ships, what may seem to all an excellent idea can result in catastrophic structural failures and oil spills.

An Imagined Use of the Paradigm

A design change to which no one seems to have objected was the cause of the structural accident that killed over one hundred people in the Kansas City Hyatt Regency Hotel in 1981 (see Marshall et al., 1982; Petroski, 1985). Here the analogy with Galileo's paradigm is less visual than conceptual, but there are nevertheless strong parallels between the cases.

The original conceptual design of the elevated-walkway support system consisted of long rods attached to the roof and passing through box beams under the fourth-floor walkway and continuing down through similar box beams under the second-floor walkway, as shown in Figure 4.6. The load transmitted to the end of each box beam was to bear on a washer, which in turn rested on a nut on the threaded rod. The obvious design problem for the connection detail was to size the box beam, rod, washer, and nut to support the apportioned dead weight of the walkways and any live load that they might carry so that the nut would not slip and the rod would not break or pull through the box beam. This appears to have been done

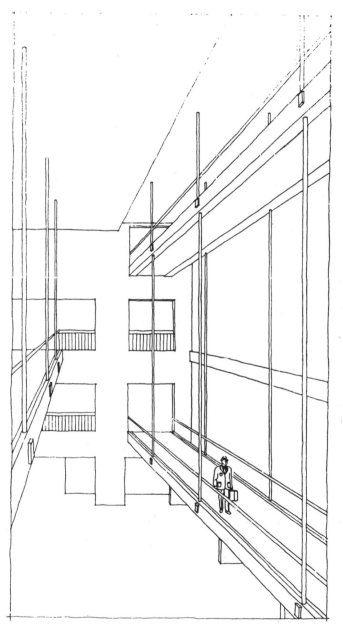

Figure 4.6. Architectural rendering of suspended walkways in the Kansas City Hyatt Regency Hotel (Marshall et al., 1982).

for the original design, as shown on the right in Figure 4.7. Although there remains some question as to whether the building code was satisfied, postaccident tests demonstrated that the connections as originally designed had a factor of safety of approximately two. In other words, the walkways should have had reserve strength equal to that required just to hold them up. Had the changes not been made and had the subsequent accident not called attention to their details, the walkways would very likely be unremarkable and safe structures still in use today.

However, a change in the walkway support was suggested and agreed to by the various parties involved. The change, which is illustrated on the left in Figure 4.7, consisted in replacing each single long rod with a pair of rods offset on the upper walkway's box beam. This made it possible to use shorter rods, which were more readily available and easier to install, and may also have avoided some confusion as to the implementation of the original design. But the new arrangement meant that the weight of not just the top but both walkways would now bear down on the washer and nut under the top box beam, thus effectively doubling the bearing strength required of the connection. And neither the box beam nor the washer and nut beneath it seems to have been resized in the changed design. Consequently, on July 17, 1981, as a group of spectators overloaded the upper walkway, the washer, nut, and rod pulled through the box beam (see Fig. 4.8) and both walkways collapsed.

Whether it was a design change of convenience or a change thought to improve safety by reducing the possibility of weakening long rods by abusing them during erection may never be known for certain, but a change for any reason must have its implications measured against the original design objectives, which in this case included the imperative ones that the rod not break or punch through the box beam. Evidently no such reconsideration took place with any sufficient care, and the walkways as built thus had a factor of safety just barely over unity against the rod being pulled through the box beam. The connections could (just barely) support the

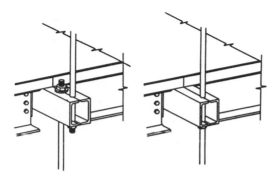

Figure 4.7. Connection detail of upper suspended walkway in the Kansas City Hyatt Regency Hotel, which failed in 1981: *left*, as built; *right*, as originally designed (Marshall et al., 1982).

weight of the walkways, but the additional weight of people watching a dance below evidently overloaded the marginal design.

A knowledge of the story of Galileo and the marble column and an appreciation of its implications for checking design changes might have alerted someone involved with the walkway redesign that here was a detail to be checked anew. Simple back-of-the-envelope calculations well within the grasp of a sophomore engineering student could have caught the error in the walkway support. But calculations are not made in the abstract; someone must be motivated to make them and must realize their relevance and significance for a possible new or altered failure mode. What makes the Kansas City walkways collapse an especially tragic case study is that the critical failure mode, that of the rod being pulled through the box beam (as shown in Fig. 4.8), was one that must have been considered at least implicitly in sizing the connection in the first place.

Conclusion

The diverse examples of ships and walkways touched upon in this chapter may be supplemented by other design scenarios in which

Figure 4.8. Failed walkway connection (Marshall et al., 1982).

the effective use of the paradigm might be imagined. The solid
booster rockets of the space shuttle, for example, were modeled after
those of the Titan III, a proven design. The Titan was a very successful rocket whose joints contained a single O-ring. In adapting
the design for shuttle use, it was argued that adding a second O-
ring (Fig. 4.9) would make the design even more reliable (see, e.g.,
Bell and Esch, 1987; Mark, 1987). This logic was evidently agreed
to by all concerned, and the belief that the double O-ring joint was
so robust and dependable may have contributed to the decision to
launch *Challenger* on January 28, 1986, under ill-advised conditions
(cf. U.S. Presidential Commission, 1986). Ironically, the redesign of
the space shuttle in the wake of the *Challenger* explosion included
the introduction of a *third* O-ring in critical joints, presumably for
further increased reliability. Familiarity with a paradigm like that of
Galileo's marble column certainly should caution against believing
that more is necessarily better in structural or mechanical reliability,
and greater real caution in future designs may result because of this
caveat.

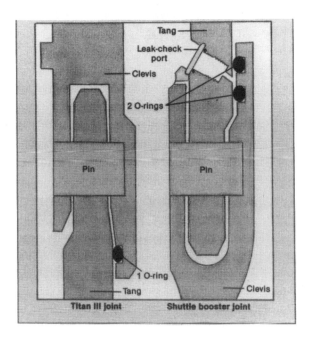

Figure 4.9. O-ring designs for Titan III and space shuttle booster rocket (Bell and Esch, 1987; by permission of The Institute of Electrical and Electronics Engineers; © 1987 IEEE).

The value of a paradigm like that of Galileo's story of the marble column is not in its literal telling but in its power to evoke analogies that lead to the drawing of generalized caveats. And if such caveats can become as familiar and intuitive to designers as are their success-based rules of thumb, then cautionary paradigms may indeed serve to reduce human error and thus increase the reliability of designs.

5

Galileo's Confirmation of a False Hypothesis

A Paradigm of Logical Error in Design

Among the most crucial assumptions in the solution of any engineering problem, whether it be a problem in engineering science or in engineering design, is the assumption of how any particular mode of failure will occur. Indeed, it is the analyst's or designer's preconceived ideas about failure that drive the analysis or design (which beyond conceptualization necessarily proceeds by some form of analysis), and virtually every theoretical calculation or experimental measurement on an analytical or scale model of a real or projected system is most significant in its relationship to how the system is imagined to fail in theory or practice.

As an illustration of these ideas, a fundamental problem that Galileo (1638) identified as central to determining the strength of materials will be explicated in this chapter. Galileo's famous problem of the cantilever beam may be taken as a paradigm of all engineering design problems, and his flawed analysis may serve as a paradigm of logical error in the solution of such problems. The unassailable genius of Galileo and the undeniable greatness of his many contributions to engineering mechanics make his error all the more suitable as the basis for a paradigm intended to emphasize the ease into which we all can fall into error.

Galileo's Problem of the Cantilever Beam

The several examples of notable failures of Renaissance engineering with which Galileo opens his treatise served for him as counterexamples to the prevailing hypothesis that geometry alone was sufficient to analyze and design structures and machines. On the contrary, Galileo asserts, geometry has to be supplemented with a knowledge of the strength of materials if future failures were to be avoided. He considers it a "fundamental fact" that all solids have a breaking point:

To grasp this more clearly, imagine a cylinder or prism, AB, made of wood or other solid coherent material. Fasten the upper end, A, so that the cylinder hangs vertically. To the lower end, B, attach the weight C. It is clear that however great they may be, the tenacity and coherence between the parts of this solid, so long as they are not infinite, can be overcome by the pull of the weight C, a weight which can be increased indefinitely until finally the solid breaks like a rope.

Galileo's illustration, here Figure 5.1, may be taken as that of a tension test designed to measure the ultimate tensile strength of the material of the specimen AB. After having spent the bulk of the first day of his dialogues on digressions "treating of the resistance which solid bodies offer to fracture," early in the second day Galileo observes that

though this resistance is very great in the case of a direct pull [as in Fig. 5.1], it is found, as a rule, to be less in the case of bending forces. Thus for example, a rod of steel or of glass will sustain a longitudinal pull of a thousand pounds while a weight of fifty pounds would be quite sufficient to break it if the rod were fastened at right angles into a vertical wall.

With this Galileo introduces the fundamental problem of the cantilever beam, which is exemplified in the famous illustration, reproduced here as Figure 5.2.

In modern terminology, the problem is essentially as follows: Assuming perfectly elastic behavior, and given that the ultimate tensile

Figure 5.1. Galileo's representation of a tensile test (Galileo, 1638).

Figure 5.2. Galileo's cantilever beam (Galileo, 1638).

strength of the material has been measured in pure tension, as in Figure 5.1, determine the load that can be carried at the end of a cantilever beam of the same material and of given cross section. (Alternatively and equivalently, the strength of material, load, its distance from the wall, and the beam's cross-sectional proportions may be specified, with the cross section to be determined.) Galileo's approach to this problem is the oldest surviving example of a rational treatment that uses more than a purely geometrical argument, and the conclusions he reaches enable him to explain a host of theretofore confusing mechanical phenomena, including the strength of levers and of simply supported beams, which may be considered to be back-to-back cantilevers each supported by its mirror image across a vertical plane of symmetry. Furthermore, understanding the principles of the strength of a uniform cantilever beam enables Galileo to ask and answer questions such as what should be the profile of a beam that everywhere has equal strength while at the same time using minimum material. This is, of course, a problem in design optimization, and it would be by the use of Galileo's results as a predictive theory that such a problem could be solved.

Galileo's Central Hypothesis and Its Apparent Confirmation

The first and fundamental proposition that Galileo sets out to establish concerns the nature of the resistance to fracture of the weightless cantilever beam with a concentrated load at its end. In order to characterize this he must, of course, make an assumption as to how the beam will fail, and then carry out his reasoning for the governing mode of failure. Although Galileo's seventeenth-century dialogues contain no diagrams of force and no equations as we know them today, his reasoning is easily demonstrated with the use of free-body diagrams and is easily translated into algebraic form. These modern aids are admittedly anachronistic, but they provide a concise means of conveying the essence of the paradigm.

Galileo states his fundamental assumption about the behavior of the cantilever beam of Figure 5.2 as follows:

It is clear that, if the cylinder breaks, fracture will occur at the point B where the edge of the mortise acts as a fulcrum for the lever BC, to which the force is applied; the thickness of the solid BA is the other arm of the lever along which is located the resistance. This resistance opposes the separation of the part BD, lying outside the wall, from that portion lying inside.

In other words, Galileo sees the cantilever being pulled apart at section AB in much the same way that the cylinder of Figure 5.1 is being pulled apart – uniformly across the section. A modern free-body diagram would show Galileo's assumption about the stress dis-

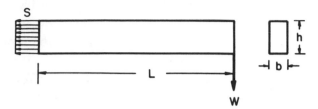

Figure 5.3. A modern representation of Galileo's fundamental assumption about the support of a cantilever beam (drawn by Fred Avent, Department of Civil and Environmental Engineering, Duke University).

tribution at the wall as in Figure 5.3, which also defines the symbols that we shall employ in our exposition.

The modern reader who has had a sophomore engineering course in strength of materials can immediately recognize Galileo's error in assuming a uniform tensile stress across the section AB, not to mention his neglect of shear altogether. It is instructive to attempt to see how the Renaissance engineering scientist could have made such an error. Holding the sharpened end of a pencil, for example, lightly between the thumb and forefinger of the left hand and pressing down on the eraser end with the right forefinger might represent the exaggerated action of the cantilever of Figure 5.2. With such a simple physical model, it is easy to see how one can identify point B as a fulcrum, for the pencil readily appears to rotate about that point. It might take a firmer grip and a rather strong push to break the pencil, but it clearly could be done. Galileo would more likely have used something like a twig or a glass rod, in fact or in a thought experiment. If he had broken a cantilevered glass rod by bending it, the fracture surfaces would have been created so quickly that any distinction between the mode of failure of the cantilever beam and the uniformly loaded cylinder would not have been apparent. And for a very brittle material such as glass, to an eye unaided by late twentieth-century fractographic techniques and unsophisticated in the topography of fracture surfaces, the broken rod would have looked much the same whether pulled apart or bent. Thus the hypothesis of uniform resistance (or stress) at the section AB could be said to have been confirmed by observation, and the analyst could proceed with confidence.

Galileo's next step is to argue that the beam would fail when the product of the uniform resistance (i.e., stress, or load per unit area, times area over which it acts) times its equivalent moment or lever arm (i.e., half the distance AB) is the same as the end force times its moment arm (i.e., the length of the beam). In algebraic terminology, using the symbols in Figure 5.3, L is the length of the beam, b and h are the width and height, respectively, of its cross section, and S is the (assumed uniform) stress on the beam at the wall; the weight W that Galileo argues could be supported by the beam of uniform resistance is:

$$W = Sbh^2 \ / \ 2L \qquad (5.1)$$

Having established his first proposition, as embodied here in Equation (5.1), Galileo could proceed to demonstrate the validity of further propositions, which in turn further confirmed the validity of the first. In his Proposition II, for example, Galileo considers the resistance to fracture of a prism or board whose width is greater than its thickness, as shown in Figure 5.4. Representing the width ac as h and the thickness bc as b, Equation (5.1) predicts the maximum weight W. When the board is oriented horizontally, however, so that the roles of h and b are interchanged, the maximum weight X that it can carry is now predicted by Equation (5.1) to be:

$$X = Shb^2 \ / \ 2L \qquad (5.2)$$

Hence the ratio of the strength of the vertically oriented board to that of the horizontally oriented one is the ratio $W : X$, that is, $h : b$. In Galileo's illustration, this ratio is approximately $5 : 1$, and the predicted strength advantage could easily have been confirmed by some simple experiments. Thus the fundamental assumption about the failure mode, which in this comparison of relative strengths would not itself have been the focus of attention, would nevertheless have been further confirmed because of the apparently correct prediction to which it led.

Galileo goes on to demonstrate the correctness of other propositions based on the reasoning embodied in Equation (5.1), many of which deal with scale effects and the relative strength of geometrically similar beams. Hence the propositions involve ratios that mask the absolute incorrectness of Equation (5.1). Even Descartes, Galileo's critical contemporary, appears to have been convinced of the validity of his assumption, because from it followed familiar results and even the then-novel one that the shape of a beam that would be everywhere of equal strength was semiparabolic (as illustrated in Fig. 5.5), a result that Descartes believed to be approximately true (Drake, 1978). But with each endorsement and with each seemingly correct proposition derived from the fundamental one, it in turn becomes more and more above questioning as to its own correctness. As the formula might be used in design with an intuitive factor of safety, it could engender further confidence. Indeed, each seemingly successful design that followed from it would confirm its truth. And the longer successful results do come of it, the less likely the false assumption embodied in a result such as Equation (5.1) will be casually uncovered. Indeed, if after a while something goes wrong with a design in which the formula was employed, the cause of the failure will be more likely sought (and probably found) elsewhere than in the "well-established" equation. This would be normal engineering science in the sense of Kuhn (1962). The books by Addis (1990) and Benvenuto (1991) place Galileo's problem in a broad historical and philosophical perspective to make this even more evident.

Correcting the Error

When Hooke's law relating force and extension in elastic materials was enunciated in the latter part of the seventeenth century, a con-

Figure 5.4. Galileo's illustration of the relative strength of the same board oriented in two different ways (Galileo, 1638).

ceptual framework was erected in which Galileo's explicit assumptions about the location of a cantilever's fulcrum and its uniform resistance to fracture (along with his implicit assumption of inextensionality of its fibers) were incompatible with contemporary understanding of how things worked. The existence of a fulcrum at point B (in Fig. 5.2) implies that the extension of the beam's fibers increases linearly from the fulcrum. Hence so must the resistance or stress. As shown in Figure 5.6, this was essentially the assumption made as late as 1729 by the French engineer Bernard Forest de Belidor, who was merely following the earlier lead of the likes of Leibniz and P. Varignon (see, e.g., Straub, 1949), and which gives the result that

$$W = Sbh^2 / 3L \qquad (5.3)$$

This is, of course, a more conservative result than Equation (5.1), but it still predicts the beam's capacity to be a full 100 percent above what it ideally would be.

As is so often the case, errors of theory are ultimately uncovered by the test of application. While designing pipelines to supply water to the palace at Versailles, Edmé Mariotte took an interest in the bending strength of beams. He experimented with wooden and glass rods and found that Galileo's theory overpredicted their breaking

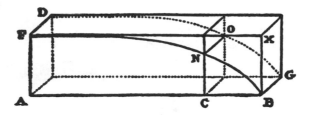

Figure 5.5. Galileo's illustration of a beam of equal resistance everywhere along its length (Galileo, 1638).

load. Mariotte conducted an elaborate series of experiments, some of which are suggested in Figure 5.7. At first he followed Galileo's idea by maintaining a fulcrum at the bottom point, but incorporated the idea of a linearly varying extension into the analysis, thus essentially subscribing to Equation (5.3), which would still overpredict the breaking load. Eventually Mariotte recognized that there must be linearly varying compression as well as tension acting across the beam's section, but a further error in calculating the resultant moment led him again to Equation (5.3), the discrepancy of which with his experimental results he attributed to a "time effect" (Timoshenko, 1953; cf. Todhunter and Pearson, 1886).

According to Timoshenko, during the eighteenth century the majority of engineers continued to use formulas based on the erroneous work of Mariotte, even though a correct treatment of the problem was published as early as 1713 by the French mathematician and scientist A. Parent:

The cause of this may perhaps lie in the fact that Parent's principal results were not published by the [French] Academy and appeared in the volumes of his collected papers which were poorly edited and contain many misprints. Moreover, Parent was not a clear writer, and it is difficult to follow his derivations. In his writing he was very critical of the work of other investigators, and doubtless this made him unpopular with scientists of his time.

Timoshenko's analysis of the loss of Parent's work to the community of designers and analysts points out the important role that human nature and nontechnical details can play in what might ap-

Figure 5.6. A modern representation of how extensibility was incorporated into an early analysis of the cantilever beam (drawn by Fred Avent).

pear to be a purely technical question decidable on purely technical grounds. As a result of the lack of clarity, visibility, and acceptance of Parent's work, coupled with the authority of Galileo and those who followed him more in the mainstream, another sixty years passed before the improved analysis of the cantilever beam was widely understood and disseminated.

The correct formulation, and thereby the correct solution, of Galileo's problem of the cantilever beam took so long to evolve in part because of the engineering-scientific credibility of Galileo, an incontrovertible genius. Our tendency is to overlook errors in fundamental assumptions, especially when they are asserted with authority and appear to be confirmed by results derived from them. Thus the success of the result embodied here in Equation (5.1), and even as "corrected" in Equation (5.3), appeared to reconfirm the hypothesis depicted in Figure 5.3 (and later in Fig. 5.6) every time a successful prediction (design) was made from it (incorporating the customary error-masking factors of safety). Since Galileo's principal interest was in the ratio of strengths, and since his approach confirmed what according to Descartes was mostly "common knowledge" (Drake, 1978), there was little evidence to challenge the result embodied in Equation (5.1) or later in Equation (5.3).

The correct assumption about the failure load of the cantilever beam is, of course, obtained by recognizing that the resistance at the wall is not uniform or even uniformly increasing from the "fulcrum" (both of which assumptions clearly leave an unbalanced horizontal force acting on the beam), but is uniformly varying about the neutral bending axis through the center of the section, as illustrated in Fig-

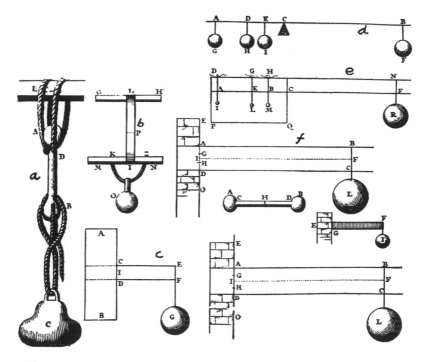

Figure 5.7. Mariotte's tensile and bending experiments (Timoshenko, 1953).

ure 5.8. The equilibrium of moments based on this diagram, which still ignores shear, leads to the correct equation of beam theory:

$$W = Sbh^2 / 6L \qquad (5.4)$$

In other words, the maximum load that could be supported by the ideal beam would be a factor of three less than what Galileo's analysis would have predicted and a factor of two less than that of Mariotte.

Could the Error Have Been Corrected Earlier?

Galileo did not write equations like (5.1), and he reasoned only about relative strengths of beams, but it is clear that his results could have

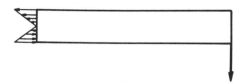

Figure 5.8. The correct assumption about bending stresses in a cantilever beam, with shear stresses not shown (drawn by Fred Avent).

been employed to *design* cantilever beams. Given a weight W to be supported at distance L from a wall by means of a beam whose tensile strength was S, the designer could size the beam's cross section by means of Galileo's reasoning as embodied in Equation (5.1) supplemented by a judgment regarding the ratio of height to width. Neglecting such effects as the statistical variation in strength of materials in order to simplify the argument, if the designer used a factor of safety greater than three, then the beam would probably have worked. And beams so designed with "Galileo's formula" might have further confirmed the correctness of the fundamental assumption about the failure mode, and thus pushed it further from the consciousness of users of the formula.

Conservatism amounting at least to the equivalent of a factor of safety of three would probably not have been unreasonable to expect in, say, Renaissance shipbuilding, where inexplicable failures were among the incidents motivating Galileo to introduce his new science of strength of materials. As late as the middle of the nineteenth century factors of safety of six and seven were commonly used for iron railroad bridges, whose technology paralleled the iron shipbuilding of the time. Thus it is very likely that any use in design of the reasoning embodied in Equation (5.1) might not lead obviously to failures.

But clearly it would have been possible to put Equation (5.1) to the test in the laboratory. Had Galileo or his contemporaries actually broken some beams under controlled conditions, they would have

found his result on average to be off by a factor of three and would
clearly have had to reexamine his reasoning, which in turn might
very well have led to an uncovering of his fundamental error. Why
was this apparently not done? After all, there was already the example of Leonardo considering a scale effect, and Galileo's appreciation of it was sophisticated enough for him to recognize the
significance of the weight of a cord or wire itself in addition to the
concentrated weight it supported (cf. Williams, 1957). It appears that
the theory embodied in Equation (5.1) was not tested in practice not
only because that was not then a prerequisite of sound engineering
science but also because the result was believed to be incontrovertibly confirmed by its ability to predict the relative behavior of boards
and beams, a throwback to purely geometrical reasoning.

Indeed, if Equation (5.1) had been used to design beams, ships,
and the like, essentially using a factor of safety significantly greater
than three, then such structures would no doubt have worked. However, as appears to be the case in engineering design generally, the
very success of those structures would have led to reductions of
conservatism in subsequent structures of a similar kind, with the
factor of safety, whether or not explicitly employed, inexorably approaching three from above. As this phenomenon progressed, structural failures would have begun to occur. At first these failures might
have been attributed to poor materials or worksmanship, or to im-
proper use or maintenance, thereby not calling into question the
validity of the result embodied in Equation (5.1). But if the factor
of safety had not then been sufficiently increased, to allow for material variation and human error, failures would have no doubt continued to mount, even as materials, workmen, use, and maintenance
were more closely monitored. Eventually, as the scenario appears
naturally to go, a very visible or costly failure would have demanded
a careful and thorough reexamination of the state of the art, which
might finally have included a critical look at the fundamental assumptions on which it, as embodied in the likes of Equation (5.1),
was based.

The Value of the Paradigm

The story of Galileo and the cantilever beam is of value to modern engineering science and design not merely as a historical anecdote but as a paradigm of error that has been repeated throughout the history of engineering. Engineering education and practice that ignore the process whereby formulas are derived and used can be as wanting as those that ignore the rules of calculus and the laws of nature. Engineering is a human activity, and thus it is subject to human fallibility and to human nature. These human traits left unexamined in the engineering process can be its weakest links. The most assiduous but uncritical application of a fundamentally incorrect analytical assumption or design formula can be expected not only to lead ultimately to design failures, but also to prolong the incorrect use of the formula if analysts and designers look everywhere but to the fundamental formula for the causes of the failures.

The de Havilland Comet was developed by British aircraft engineers out of their experience with World War II fighter jet aircraft. It was to be the first passenger jet airliner to make regular transatlantic flights, and its inaugural year of service (1952) appeared to confirm the success of its design. Soon, however, some inexplicable midair explosions of Comets taking off from Calcutta and Rome proved troubling. Weather and pilot error were blamed at first, and subsequent accidents prompted all sorts of design changes. But one fundamental design assumption, that metal fatigue would *not* be a determining factor in the structural integrity of the aircraft, was repeatedly reaffirmed as correct – except by one maverick researcher who secured a Comet and carried out full-scale cyclic loading tests on it at the Farnborough Royal Aircraft Establishment, in England, until the plane's pressurized fuselage failed catastrophically from metal fatigue (see, e.g., Petroski, 1985).

The Hartford (Connecticut) Civic Center roof was an ambitious space frame that was designed with the aid of an elaborate computer model. The roof covered two and a half acres of an arena from which thousands of basketball fans had exited only hours before the entire roof collapsed in January 1978. Evidently, ice and snow had overloaded the structure and caused it to fail by the successive buckling of struts that allowed the entire roof to fold into the arena. Postaccident studies revealed that incorporated into the computer model were some fundamental assumptions about the end conditions on thirty-foot-long members of the frame, assumptions that proved to be grossly oversimplified. But so much confidence was apparently placed in the supposedly sophisticated computer model that even some reportedly alarming deflections and flexibility in the roof during construction did not cause a sufficiently critical reexamination of the fundamental assumptions in the model. Each day that the roof had stood was of course taken as further confirmation and vindication of the computer model. But the incontrovertible evidence of the catastrophic failure was the only counterexample needed to disprove the infallibility of the fundamental hypothesis that the model was correct (see, e.g., Petroski, 1985; Zetlin Associates, 1978).

A less dramatic example of an error that might be classified under the Galilean paradigm occurs in the recent literature of applied mechanics. It involves the prediction of the fracture of pencil points (see Cowin, 1983; Cronquist, 1979; Petroski, 1987b; Petroski, 1990). Though not claiming lives and with consequences certainly not of the magnitude of most errors in analysis, this example points out how analysis can be hampered by uncritically accepted first assumptions. A pencil point in use is essentially an obliquely loaded cantilever beam; the problem in question is essentially to predict where and how such a point would break annoyingly under the pressure of a writing deadline.

Like Galileo's first attempt to analyze the classic cantilever, the first recorded attempt to analyze the fracture of pencil points (Cronquist, 1979) was based on the assumption that failure would occur where the axial tensile stress exceeded the ultimate tensile strength of the brittle ceramic pencil lead. Using tools no more sophisticated than elementary strength of materials and calculus, Cronquist was able to predict the size of broken-off pencil points, thus confirming his failure hypothesis, but he admittedly could not explain why the points had a slanted fracture surface, as suggested in Figure 5.9. The basic failure criterion (maximum tensile stress) of this first analytical attempt was confirmed as a reasonable fracture criterion when a fairly accurate prediction of broken-point sizes was achieved. Indeed, this was taken as a confirmation of the basic assumptions of the simple analysis, and more accuracy was expected from a more sophisticated treatment of tensile stress. However, while this failure criterion is relevant to the *initiation* of the fracture, it does not alone govern the continued propagation of the crack through the pencil lead, which is affected by shear as well as by tensile stresses (cf. Walker, 1979), a point overlooked in early treatments of the problem. This simple observation, so obvious in retrospect, illustrates how convincing confirming predictions can be that a hypothesis is correct and complete. The pencil point problem is trivial by engineering standards, but it is a clear example of the pitfalls of analysis in service to design that have always been the bane of engineers and engineering scientists.

Familiarity with a story like that of Galileo and the cantilever beam, especially with an emphasis on the overarching theme of how even a Renaissance genius can make what in retrospect is an elementary and obvious error in a novel analysis, can serve engineering students and practitioners well by providing a memorable caveat. As analysis and design proceed from fundamental assumptions through formulas and specifications to the construction and use of artifacts, our self-critical abilities can become overrelaxed as successful results appear to confirm whatever fundamental hypotheses were made at the outset.

No hypothesis can ever be proved incontrovertibly, no matter how many myriad successful designs may be derived from it. But it takes only one failure (in analysis or reality) to provide a counterexample to a hypothesis, and it is the engineer's responsibility to recognize this. While not every failure is conclusive proof that there are fun-

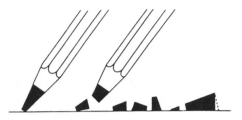

Figure 5.9. Broken pencil points showing slanted fracture surfaces (drawn by Fred Avent).

damental flaws in the state of the art, it should be seen as a professional responsibility of the engineer to recognize an obligation to the engineering method to look critically at every aspect of the design and manufacturing process that went into creating the failed artifact, including the most basic assumptions of design and analysis. Should those assumptions be flawed but not identified as such, all design "corrections" or refinements could be for naught.

6

The Design and Collapse of the Dee Bridge

A Paradigm of Success Masking Error

Except in cases where spectacular public works or loss of life is involved, the details of specific engineering failures are seldom discussed outside the narrow discipline of obvious relevance. Recently, there have been some notable exceptions, of course, including the collapse of the elevated walkways in the Kansas City Hyatt Regency Hotel in 1981 and the explosion of the space shuttle *Challenger* in 1986. Such colossal failures have been headline news, and few engineers of any discipline are unaware of the most probable causes of the failures, at least in a general sense.

Less spectacular and less recent failures tend to be remembered only within the industry or discipline for which they are perceived to have a continuing relevance, and even then they tend to be relegated to dusty archives as the state of the art appears to eclipse the errant technology that was involved. But a considerable number of failures throughout engineering history have been due to errors in the same timeless design logic and methodology that are used today, and so the root causes of classic failures can and do have a continuing relevance for current designs and design processes of the greatest sophistication and complexity. Furthermore, because lessons drawn from errors in the design process itself necessarily transcend the specific flawed application, failure case histories that are employed as vehicles for conveying the nature of a *type* of design error have significance for the design process generally and timelessly.

Case histories of failures often tend to focus in excruciating detail on the specifics of the failed design and on the failure mechanism. Such a concentrated view often discourages all but the most narrow of specialists from pursuing the case history to any degree, for the lessons learned and conclusions drawn tend to be so case-specific as to seem hardly relevant to anything but a clone of the failed design. There have been some notable exceptions, of course, including the report of the Presidential Commission that investigated the *Challenger* accident. Not only was the root technical cause of the problem elucidated but also the important role of the management–engineer interface and its pitfalls was explored in considerable detail. However, the volumes of background, evidence, and analysis, not to mention the social, political, and technical complexity of the case, make it a diffuse one from which to convey deeper lessons concisely.

The collapse of the Tacoma Narrows Bridge presents a more purely technical case history, but because filmed images of it have been so widely distributed and because so many imprecise and incomplete explanations of the physical cause of the failure have been disseminated, discussions of the collapse tend to be charged with prejudgment and oversimplification. As late as 1990, the fiftieth anniversary of the bridge's collapse, papers in scholarly journals disagreed as to what really caused the failure (cf. Billah and Scanlan, 1991; Petroski, 1991d).

There are, however, case histories of engineering failures that do not suffer from overexposure, overanalysis, or overly charged debate. These are examples from simpler times and simpler engineering, from which can be distilled more easily lessons about elements of the design process itself. Because these cases are less well known, rest upon analytical foundations that are no longer matters of debate, and involve political and social components of another era, the case histories themselves can be presented and read more for their lessons about the nature of the design process itself rather than for an account of its manifestation in a particular artifact. The Dee Bridge is one such case history.

The Dee Bridge as a Paradigm

Sufficient background on the Dee Bridge will first be provided in order to put the history of the bridge and its design in a contemporary context, but the main thrust of this chapter will be toward drawing lessons about what it was about the design process itself that allowed a critical aspect of the design to go unnoticed and thus become the root cause of the bridge's failure. The critical flaw in the human design process that produced the ill-fated Dee Bridge is paradigmatic; the same fundamental flaw was rooted in the design process and environment that produced the Tacoma Narrows Bridge almost a century later (as well as a host of other doomed bridge designs). The lessons of the Dee Bridge have a relevance well beyond the design of bridges, however, and hence the case history is presented as a paradigm for that kind of human error in design that may be termed the success syndrome.

The Dee Bridge was designed in an environment of success and confidence, for many bridges like it had been erected over the previous decade and a half and had been performing well both structurally and economically. What distinguished the Dee from its predecessor designs was that it was geometrically at the outer limits of experience with the design class (see Fig. 6.1), which itself had evolved some novel features that distinguished in rather subtle ways newer examples from older ones. Furthermore, no doubt as a result of then-recent successful experience, the latest design had less rigidity and a lower factor of safety than its predecessors (see Fig. 6.2). The presence of all these elements contributed to causing a failure mode that had been latent and therefore of insignificant consequence

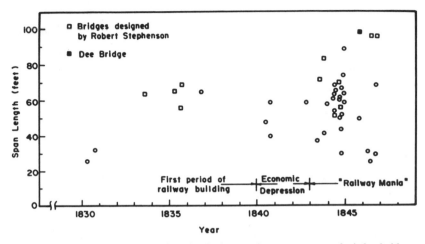

Figure 6.1. The building and length of nineteenth-century trussed-girder bridges (after Sibly, 1977).

in the design class finally to dominate the behavior of the Dee Bridge.

Design of the Dee Bridge

Like all design problems, bridging the Dee River at Chester, in the northwestern English county of Cheshire, had no unique solution, and it was the British engineer Robert Stephenson's role to decide what kind of bridge provided a satisfactory compromise to the often competing constraints that included anticipated rail traffic, site conditions, and economic considerations of the 1840s. At first Stephenson conceived of a five-span bridge of brick arches, but riverbed conditions could not support such a massive structure. A timber viaduct would have been light enough, but questions of rigidity and durability, not to mention the scarcity of timber in Britain in the mid-1840s, argued against this option. Thus an iron bridge of some form appeared to be a logical choice, and a girder bridge would be among the lightest and therefore most economical. The story of the

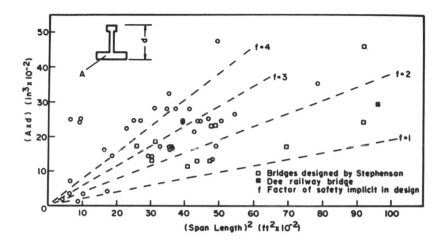

Figure 6.2. Factors of safety of trussed-girder bridges (after Sibly, 1977).

Dee Bridge has been told in some detail by Sibly (1977) and Beckett (1984), among others, and their reports provide much of the detail for and a great deal of amplification on the following account of the design and failure of the bridge.

Once Stephenson decided upon a girder bridge, the girders themselves had to be sized. Cast iron had long been known to be much weaker in tension than in compression, and so, although it was a suitable material for an arch structure like Iron Bridge (see Figure 7.3), erected over the Severn River in western England in 1779, it was a poor choice for structural members loaded transversely the way a girder was. Nevertheless, in the first part of the nineteenth century iron beams were being cast in a variety of forms having a much greater concentration of material on the tension side so that bending action could be resisted in a more or less economical way. By the early 1830s, the engineers William Fairbairn and Eaton Hodgkinson had tested a variety of cast-iron beams in search of an efficient, if not optimal, cross section, and in 1831 Hodgkinson published an empirical formula relating the concentrated centrally applied load that caused failure, W, to the area of the bottom flange,

A, the depth of the beam, *d,* and the span between simple supports, *L.* With lengths measured in inches, the formula yielded the failure load in tons:

$$W = 26Ad \ / \ L \qquad \qquad (6.1)$$

Hodgkinson's formula does not specify anything about the area of the top (compression) flange and does not, of course, describe the actual shape of the beam. What the formula did do was give designers like Stephenson an analytical means of sizing cast-iron beams of a relatively predetermined cross-sectional geometry so that their tensile strength was nowhere exceeded. The top flange of beams designed by the Hodgkinson formula was logically very thin and narrow compared to the bottom, with the two flanges sized relative to each other so that their areas were in inverse proportion to the relative strengths of cast iron in compression and tension, which are about 16 : 3. The form of the girder that Stephenson settled upon for the Dee Bridge, for example, is shown in Figure 6.3, and the areas of its top and bottom flanges are exactly in the ratio 3 : 16. Thus, in the elementary analysis of the time, the Dee Bridge girders under a central load resisted both tension and compression in an optimal way.

The weight of the girders themselves was distributed over the entire length of the span, of course, but the location of the load of a railroad train would naturally vary as it crossed the bridge. Such conditions were taken into account by introducing worst-case or equivalent central loads. For example, a uniformly distributed load of $2W$ would be expected to be required to break the beam characterized by Equation (6.1).

By elementary calculations, Stephenson could establish that sizing a girder such as in Figure 6.3 would give it a factor of safety of about one and one-half against the expected load of a railroad train (cf. Fig. 6.2). This meant, of course, that the theoretical breaking load of the girder was only about 50 percent greater than the maximum load it was expected to carry. But all such calculations took

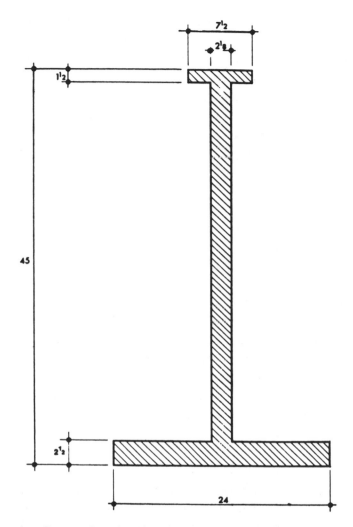

Figure 6.3. Cross section of a Dee Bridge girder, with dimensions in the contemporary units of inches (Beckett, 1984; by permission of David & Charles, Newton Abbot, Devon, England).

into account only static loading conditions, and the dynamic loadings of railroad trains were expected to be taken into account with a sufficiently large safety factor, certainly one greater than one and one-half. Structural designers of the time preferred a safety factor in the range of at least two to four, and often as high as six or seven, but Stephenson's inclinations were toward the lower range of safety factors for the beam proper, as shown by Sibly (1977; see also Sibly and Walker, 1977). He thus sought other ways to increase the margin of safety for the Dee Bridge.

Rather than resize the beam to larger dimensions, which in turn would naturally have also increased its weight, Stephenson chose to employ a technique he had adopted for several other railway bridges over the preceding decade. The trussed girder was introduced for spans of the order of thirty feet as early as 1831 by the British civil engineer Charles Blacker Vignoles. The composite structural system employed wrought-iron bars essentially to prestress a segmented cast-iron beam, thus taking some of the tensile load off the bottom flange at the beam's midspan. The components of a trussed girder are illustrated in Figure 6.4, with the humpbacked elements providing the means of connecting beam segments and attaching the wrought-iron tension bars. Stephenson and other engineers employed the principle in increasingly longer spans, as shown by Sibly (1977) and Sibly and Walker (1977), but before the Dee Bridge application no trussed girder longer than eighty-eight feet had been employed. The Dee was designed with three trussed-girder spans of ninety-eight feet each. Furthermore, most earlier bridge designs had not relied upon the wrought-iron ties for any fundamental structural strength; they were intended to keep the bridge tied together for temporary service should the brittle cast-iron beam fracture for any reason. However, Stephenson came to rely more and more on the ties for prestressing the beam and thus carrying some of the tensile load under normal service.

As discussed by Sibly (1977) and indicated in Chapter 3, the application of scaled-up trussed girders to ever-longer bridge spans

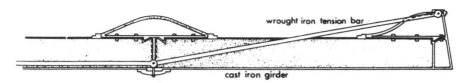

Figure 6.4. The longitudinal structure of a trussed girder (Beckett, 1984; by permission of David & Charles, Newton Abbot, Devon, England).

should have been accompanied by extreme attention to and careful observation of their behavior. The first trussed girders, being of an innovative and thus not wholly tried nature, would naturally have been overdesigned by their inventor Vignoles, in part to obviate reasonable objections to them. He may have been especially conscious and hence cautious of the fact that the tension in the wrought-iron bars was a matter of assembly and not of calculation, for the structure was statically indeterminate. Since the principles upon which Vignoles' invention was based would not likely have been easily articulated to answer any and all critics, the girders most likely would have had to have been sufficiently stiff (even without the trussing) to be convincing in a proof test (cf. Petroski, 1992b). However, mechanical and other phenomena of no significant consequence in the overdesigned thirty-foot spans would grow in relative importance as the spans of application were increased, while at the same time the proven success of shorter trussed girders in actual structures would have shifted the focus of any debate away from a need for designers to prove safety and toward a need for critics to prove risk. Without the benefit of statically determinate calculations, warnings of risk would have been as hard objectively to justify as assurances of safety.

The Accident

The Dee Bridge was completed in September 1846 and opened to railroad traffic the following month. The bridge exhibited noticeable vibration under passing trains, but the absence of trouble with other trussed-girder structures forestalled closer scrutiny. In May 1847, five inches of ballast was added to the roadway (see Fig. 6.5) to cover the timber planking on which the rails were laid. This action was taken at least in part to reduce the risk of a fire being touched off by hot coals falling from a passing locomotive, and the addition of the ballast may have been expected to dampen some of the bridge's vibrations. Stephenson personally supervised the roadwork, and he appears to have been satisfied that the added weight (without structural strengthening) did not reduce the factor of safety below prudent limits. Evidently it did, however, for the driver of the first train to cross the newly ballasted bridge felt it "sinking" under him and opened the throttle to rush across. Only the locomotive reached safety, and the five carriages it had been pulling crashed into the river with the bridge span. An engraving of the scene of the failure is shown in Figure 6.6. The two-track bridge's trussed-girder construction is clearly visible on the track that did not collapse, with the humplike connection pieces between girder segments giving the structure its characteristic profile.

Failure Analysis

The most likely cause of failure was a torsional buckling instability to which the bridge girders were predisposed by the compressive loads introduced by the eccentric diagonal tie rods on the girder. Since there was essentially no lateral bracing, the six- to seven-foot depth of the narrow-topped girder with its protrusions proved to be unstable at the load level created by the additional ballast, which also may have exerted some amount of transverse force at track level. Additional tension in the wrought-iron tie rods, due to the added ballast, naturally induced additional compression into the segmented beam, which was inclined to twist out of its plane by the manner in which the track load was transferred through the bottom flange of

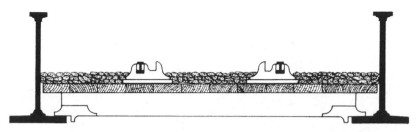

Figure 6.5. Cross section of one of the Dee Bridge trackways, showing ballast added above the wooden decking (Beckett, 1984; by permission of David & Charles, Newton Abbot, Devon, England).

the girders. The additional weight and vibrations superimposed by the passing train apparently proved to be enough of an incremental disturbance to trigger the latent torsional instability and throw the bridge and its burden into the Dee.

The failure of the bridge killed five people and injured another eighteen. The loss of life called for a coroner's inquest, which in turn led to the recommendation that a Royal Commission be appointed to look into the application of iron in railway structures. This occasioned, among other things, the first detailed but misdirected discussion of metal fatigue (then believed to be due to some mysterious vibratory mechanism that transformed the metal's granular structure).

Stephenson argued that the failure was caused by a derailment that in turn broke out a piece of the (brittle) girder, but his explanation convinced few of the investigators at the inquest. Testimony from contemporary engineers was sought, and to ascertain what factor of safety was used in common practice they were asked the question, "What multiple of the greatest load do you consider the breaking weight of the girder ought to be?" The responses ranged from three to seven. Stephenson was, of course, using a factor well below that, and the commission concluded that for future designs a prudent factor would be six.

Among those testifying before the Royal Commission was the fa-

Figure 6.6. The collapsed Dee Bridge (*Illustrated London News*, June 12, 1847).

mous British engineer of the Great Western Railway, Isambard Kingdom Brunel (cf. Petroski, 1992a), who argued against rigid rules (what today would be called codes) for bridge building. He went so far as to refer to the investigating body as "the commission for stopping further improvements in bridge building" and believed that, with the proper care in eliminating inhomogeneities and other imperfections, reliable iron castings "of almost any form and of twenty or thirty tons weight" could be ensured. Brunel, like Stephenson, seemed unaware of, or unwilling to acknowledge, the potential dangers of physical phenomena that do not manifest themselves until successful systems are scaled up in size or trimmed down in strength to the point where these phenomena, once dominated themselves, become dominant in limiting the system's performance. No matter how homogeneous and free of imperfections the components of a trussed girder might be, for example, there will be a size of it whereby torsional instability is a much more serious concern than tensile overload. When, because of ignorance of torsional instability, tensile overload becomes the dominant design consideration, there is naturally a size of girder that will be unsafe. That size will be larger the larger the factor of safety, of course, but a limiting size will nonetheless exist. Stephenson found the limits of his design in the Dee Bridge girders.

Lessons for Design

Virtually all design is conducted in a state of relative ignorance of the full behavior of the system being designed. While the example of the Dee Bridge involves only mechanical phenomena, the implications of it as a case study are valid *a fortiori* for systems involving the complex and coupled interactions of biological, chemical, electrical, nuclear, thermal, and other effects. As long as there remain open problems in the engineering sciences, there will be insufficient calculational and predictive theoretical tools to design without the factors of safety that are so frequently and correctly also called "factors of ignorance." The advent of supercomputers has facilitated complex computations, but their results are only as complete as the theories on which their programs are based. And even the most mathematically rigorous of theories are only as complete as the physical understanding on which they are based and interpreted.

Although mathematicians and scientists do not always like to be reminded of the fact, a good deal of our technology leads their understanding of theories behind its success. The classic example is, of course, the steam engine, which was invented and developed to a high degree of reliability long before there was an engineering science of thermodynamics. Indeed, it was the very artifact of the working steam engine that called out for a theory of its operation. Likewise, the airplane was developed before a sophisticated theory of flight, and certainly before the now-indispensable theories of aer-

odynamics and aeroelasticity were able to explain such disturbing
phenomena as stall and flutter and thus be called into engineering
service to troubleshoot and optimize designs. In fact, one can go so
far as to say that the ever-evolving world of human-designed artifacts
can always be expected to hold challenges for the engineering scientist no less than does the natural universe hold challenges for the
natural and physical scientist.

Invention and design can proceed with a qualitative ingenuity that
gives us things that work. The ability of the inventor and designer,
in concert with the artisan and craftsperson, to produce first working
models and finally prototypes can rest entirely upon experience and
an empirical sense. Sketching, sizing, making, testing, and debug-
ging are design activities that can proceed with little or no theory,
and they can proceed from failure to failure to success. It is the part
that does not work that gets the attention in the model and the
prototype. It is the function that does not get executed as well as
can be imagined that becomes the focus of improvements. In time,
the design process yields an artifact in which failure is believed to
have been anticipated and obviated to the extent required for the
application. However, there can always be latent physical and other
phenomena that, while bearing little relevance for the initial artifact,
can prove to dominate the behavior of scaled-up or pared-down
descendants.

It is a truism to say that the application of all designs is a matter
of future use. Thus the designer who does not properly anticipate
future use and behavior is liable to being surprised by the design.
Among the most common design problems is the scaling up of existing designs – making a more powerful engine or a longer bridge
– and the corollary weakening of successful designs. Ironically, such
activity appears at the same time to be what might be termed safe
and (after Kuhn) normal design, which presents little but an incremental challenge to the designer (and holds little interest for the
analyst), and what might be called daring design, in the sense that
it goes beyond the envelope of experience and thus risks encounter-

ing unknown phenomena that may dominate the design's behavior. Because the safe and normal aspect of the design activity takes place in the context of a record of successes, the element of risk tends to be minimized if not completely ignored. Critics, having no more convincing theory than proponents of the design extrapolation, cannot come up with incontrovertible arguments that failure is just around the design corner.

The Dee Bridge failed because torsional instability was a failure mode safely ignored in shorter, stubbier girders whose tie rods exerted insufficient tension to induce the instability that would become predominant in longer, slenderer girders prestressed to a considerable compression by the trussing action of the tie rods. Similarly in the twentieth century, a dynamic torsional instability was safely ignored in heavy suspension bridges with relatively wide and deep deck structures, but it proved ultimately to dominate the behavior of the light, only two-lane-wide and very shallow deck of the Tacoma Narrows Bridge. The spontaneous collapse of the Quebec Bridge while the daringly slender 1,800-foot-span cantilever structure was under construction in 1907 (see Figs. 3.4 and 3.5) revealed a designed-in propensity toward a buckling instability; this instability was not at all an issue in the massive cantilever bridge (with two 1,710-foot spans) across the Firth of Forth (see Fig. 6.7) completed in 1890 near Edinburgh, Scotland, from which the Quebec descended. While both bridges were of the same scale (cf. Figs. 3.4 and 6.7), the success of the "overdesigned" Forth evidently led to an overconfidence manifested in the "underdesigned" Quebec. Gross size alone is not what brings great structures down.

It is not only increases in scale or slenderness that go beyond design envelopes, as the use of stainless steel in novel fast-neutron reactor environments demonstrated. Here, experience with light-water nuclear reactors reinforced the design paradigm that material behavior relating to neutron moderation, mechanical strength, thermal creep, and irradiation embrittlement should dominate design choices. However, the use of "properly designed" stainless steel fuel

Figure 6.7. The Firth of Forth Bridge, completed in 1890.

canisters in the accelerated neutron flux environment of fast reactors soon revealed the previously unknown material property that manifested itself in dramatic volumetric swelling. The swelling of the stainless steel canisters in the reactor core threatened to lock them in place and thus severely limited the time they could be left in the reactor between refueling operations. The need to remove and replace the swelled fuel canisters substantially affected the operating characteristics of the reactor and thus limited its efficiency. This is an example of a design surprise and functional failure that, while not catastrophic, had a limiting effect on the design's future.

Conclusion

Normal design will no doubt continue to be a dominant human activity, but it behooves designers to proceed with a heightened awareness of the human error pitfalls that seem almost invariably to lie somewhere beyond the design envelope. This is not to say that design envelopes should not be responsibly pushed outward and extended, for to do that is as much a part of being human as is error. A common heritage of classic case histories of failures, especially those that highlight the limitations of normal design and that stress the kinds of error that are so easy to make and so difficult to catch, may be the single most promising way to minimize repeating the mistakes in design logic that have plagued designers from the beginnings of time. For example, if there were a more heightened awareness among designers that in case after case throughout history there have been surprises in extrapolatory design, then designers engaged in such design situations might pay more attention to the warning signs that seem invariably to prefigure failures. Such signs appear to be ignored when the design environment rests upon an accumulation of successes rather than upon a broad-ranging awareness of failure case studies that point to common errors and lapses in design attention.

Robert Stephenson, perhaps sensitized by his experiences with the Dee Bridge, became a strong advocate of discussing failures in the engineering literature. In 1856, writing to a journal editor about a manuscript under review, Stephenson expressed the hope that in revising the paper its author would note "all the casualties and accidents" that occurred during the progress of the engineering project discussed. Stephenson had become a confirmed believer in the value of case histories, and (as quoted in Whyte, 1975) wrote that "a faithful account of those accidents, and of the means by which the consequences were met, was really more valuable than a description of the most successful works."

Another nineteenth-century bridge designer, the American George Thomson, noted in 1888 that the large number of failures of railroad bridges in the United States and Canada (over 250 in the previous decade) indicated a pathological condition among mechanical structures that deserved as much attention among engineers as medical pathologies get among physicians. He argued his analogy further:

The subject of mechanical pathology is relatively as legitimate and important a study to the engineer as medical pathology is to the physician. While we expect the physician to be familiar with physiology, without pathology he would be of little use to his fellow-men, and it [is] as much within the province of the engineer to investigate causes, study symptoms, and find remedies for mechanical failures as it is "to direct the sources of power in nature for the use and convenience of man."

Thomson's quotation from the first formal definition of civil engineering (see, e.g., Watson, 1988) evoked the sense among early Victorian engineers that their activities were as much in the service of mankind as those of the more established professions. That sense evolved into a kind of hubris by the late nineteenth and early twentieth century, however, and the increasingly dramatic successes of engineers drove from the literature and, apparently, the mind of the engineering profession the failures and mistakes they then seemed doomed to repeat. Judging from the paucity of professional literature on failures (as opposed to failures themselves) in the first three-quarters of the present century, it was unfashionable at best, and unprofessional at worst, to deal too explicitly with the errors of engineering. This was unfortunate, but it itself appears to be an error that is nowadays being rectified.

7

The Britannia Tubular Bridge

A Paradigm of Tunnel Vision in Design

Not all engineering failures take place suddenly and dramatically, accompanied by the crack and crash of steel and the tragic loss of life. Indeed, some classic errors in design have been all but ignored and forgotten in the context of what are generally hailed as tremendously successful projects. Yet these errors can be no less endemic in the design process, and hence case studies elucidating them can be extremely instructive and valuable for the teaching and practice of design. One class of such errors may be described by the rubric of tunnel vision in design, and like other types of errors, it is best introduced through a paradigmatic case history.

One of the most fruitful examples of the role of failure in the history of engineering design was hailed as a tremendous structural success and stood for 120 years as a monument to its engineer. The Britannia Bridge was to carry the Chester and Holyhead Railway across the Menai Strait, a strategic stretch of water with tricky currents between the northwestern coast of Wales and the Isle of Anglesey. On the other side of Anglesey, at the port of Holyhead, trains from London could meet ferry boats to Dublin. Because the Menai Strait was so important to the British Navy, the Admiralty would

Figure 7.1. Commemorative painting by John Lucas of a meeting on the progress of the Britannia Bridge, showing Robert Stephenson (seated, center) and Isambard Kingdom Brunel (seated, far right) with the bridge under construction in the background (Institution of Civil Engineers).

allow no bridge that would interfere with shipping during or after construction.

Robert Stephenson's design solution was the high-level crossing to be known as the Britannia Bridge. It was generally regarded as the most significant construction project under way in the British Isles, if not the world, in the late 1840s, and engineers and designers made a point of visiting the site to see its progress. A famous painting by the artist John Lucas shows Stephenson in a meeting with other engineers and contractors, with the partially completed Britannia Bridge in the background (Fig. 7.1).

What makes the Britannia Bridge so valuable as a case study is the multifaceted way in which it points out the important role that failure plays in design. Conscious and explicit considerations of structural failure were present (and well documented) throughout the entire design process, and this fact ensured the unquestioned structural success of the revolutionary bridge design. However, the tunnel vision created by so narrow a focus on the dominant structural problem of bridging the Menai Strait resulted in a monument to engineering shortsightedness. The structural principles involved in the Britannia Bridge had a signal influence on the evolution of structures in the latter part of the nineteenth century, affecting the design of everything from cranes to ship hulls (cf. Rosenberg and Vincenti, 1978); nevertheless, the bridge proved in retrospect to be such an economic and environmental failure that its own revolutionary form was obsolete almost immediately upon completion, as evidenced by the fact that only a few bridges like it were ever built.

How and why the Britannia Bridge could at the same time be both a tremendous success and a failure as a model for later bridges is best understood in the context of circumstances surrounding the bridge's design and construction. As with all case histories meant to serve as paradigms, its story takes on greater significance as more and more details are known. The following explication gives only some bare essentials of the case history, emphasizing in particular the role of failure in the design process; considerably more detail can be found in the invaluable contemporary account by the bridge's resident engineer Edwin Clark, "published with the sanction, and under the supervision of Robert Stephenson" in 1850 in two volumes plus a folio of plates. Some introductory observations on the history of the design by Stephenson himself are included in the work. Another contemporary account, by William Fairbairn, who conducted the crucial model experiments that helped determine the final configuration of the bridge's great wrought-iron tubes, was published in 1849 and reproduces many of the letters that Fairbairn and Stephenson exchanged as the design evolved. A modern interpretation of the bridge's significance for the history of technology is contained in the monograph jointly authored by the economist Nathan Rosenberg and the engineer Walter Vincenti (1978).

On the Introduction of Iron into Bridge Building

The story of the Britannia Bridge is inseparable from the story of the development of the structural use of iron. What is generally regarded as the first iron bridge still spans the Severn River near Coalbrookdale, in Shropshire, England. The 100-foot cast-iron arch of Iron Bridge was completed in 1779, after its proponents spent years persuading Parliament to authorize an iron bridge and further years outmaneuvering financial backers who were inclined toward a more conservative concept employing then-traditional bridge-building materials like stone and timber. The importance of the Severn River for the local iron trade was so great, however, that there was a reluctance among all concerned to obstruct river traffic with any midstream piers or even the temporary timber falsework needed to erect a stone-arch bridge. Furthermore, the Severn Valley was the center of iron making, and there was an obvious preference for using local building materials. In the end, an iron bridge, whose large cast parts could be fitted together quickly and with the least disruption of commerce, was agreed upon.

In the 1770s there were few analytical tools for shaping or sizing the parts of an arch bridge, let alone one made of iron. Thus it was natural to proportion bridge parts of an untried material as if they were parts of a bridge made of a proven material: among the architectural sketches of proposed elevations for the bridge, there are clear attempts to make the iron bridge look as if it were made of stone or timber (see Fig. 7.2). The profile of the bridge as built (see Fig. 7.3) actually does have a strong resemblance to a Roman semicircular stone-arch bridge, and the details of the cast-iron ribs and how the parts fit together are strongly suggestive of the details of timber construction. In the final analysis, the responsibility for supplying the parts for Iron Bridge fell to the local forgemaster Abraham Darby III, whose inherited and direct experience with cast iron was unsurpassed.

Cast iron, like stone, was well suited to taking compressive loads,

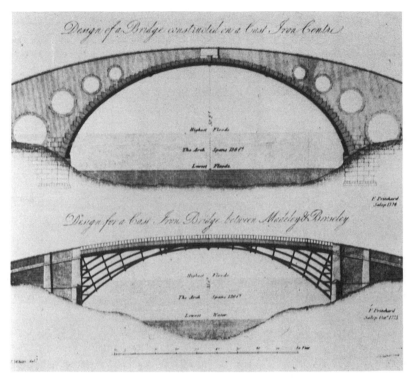

Figure 7.2. Two designs for an iron bridge across the Severn River, each mimicking construction in the traditional bridge-building materials of timber and stone (Ironbridge Gorge Museum Trust, 1979).

but it was known to be unreliable in tension, a weakness to be underscored by the collapse of the Dee Bridge. Wrought iron, on the other hand, could be forged into links and chains of great strength. Thus in the early nineteenth century it was natural to advocate the use of increasingly available wrought iron to support the roadways of bridges from above rather than from below. In this way water traffic would be virtually unimpeded not only during construction but throughout the life of the bridge. (Arch bridges did, after all, have maximum headroom only at their center.) The concept of a suspension bridge is said to have ancient Asian roots, with vines and

Figure 7.3. Iron Bridge, erected in 1779 (Ironbridge Gorge Museum Trust, 1979).

other forms of vegetation first being employed in the principal structural members. A revitalization and extension of the form took place
in the West in the early nineteenth century through the use of
wrought-iron chains.

Among the most celebrated of wrought-iron suspension bridges
was the one completed across the Menai Strait in the mid-1820s (see
Fig. 7.4). Its designer, Thomas Telford, began his career as a stonemason, and the stonework of the approaches and towers of the Menai
Strait Suspension Bridge, which still support and grace the structure, are among the features of the bridge that make it one of the
most attractive in the world. But it was Telford the engineer and
not the stonemason who had to wrestle with the hard questions of
how heavy a chain of flat wrought-iron eye-bars would be needed
to support the then-enormous span of 580 feet between the towers.
Typical of a new engineering endeavor, it was the failure character-

Figure 7.4. Thomas Telford's Menai Strait Suspension Bridge, completed in 1825 and opened early in 1826.

istics of the chain that the engineer sought to understand in order to know how to avoid failure of the structure itself (cf. Paxton, 1979).

Telford suspended a specimen chain between test towers as much as 900 feet apart and hung weights from it. He employed an equation for the catenary developed by the British mathematician, physicist, and optician Peter Barlow and observed the stretch of the chain as a function of the load it bore. Telford noted that the iron chain began to stretch more rapidly after a certain level of strain was reached, and this occurred at about half the load that the fully stretched cable could support. He therefore incorporated a sufficient amount of iron in the bridge chains so that under the worst load that he could imagine on the roadway the strain in the cables would not exceed two-thirds of the critical strain. Thus he employed a factor of safety (always an explicit acknowledgment that failure is a limiting design criterion) of one and one-half against yielding and three against ultimate failure of the chains. The most novel aspect of the design, the iron chains, naturally caused Telford to look with what proved to be tunnel vision at problems the chains themselves held for a successful structure. His concern with the failure of the wrought-iron chains was so great, in fact, that he neglected to pay sufficient attention to how the relatively light and flexible roadway might behave in the wind, and that in fact was to prove to be the bridge's weakest link.

In the meantime, the price of wrought iron had dropped dramatically. According to an observer (quoted in Hunt, 1851) writing at midcentury and looking at the development of bridge building in perspective:

It is a distinguishing element of the engineer's art, to adopt the material best suited, economically speaking, to the work he has to accomplish.

In 1806 the price of bar iron, larger size, was twenty pounds per ton; in 1816 it was ten pounds per ton; in 1828 it was eight pounds per ton; and in 1831 it was five pounds to six pounds per ton.

Thus, this material has gradually come into the domain of applications in construction, from which its high price had long excluded the consideration of its qualifications.

In such a climate, it is no wonder that iron bridges came to be the structures of choice for the expanding railroads in Britain. But a period of economic depression hit Britain in the 1830s, and few bridges, innovative or otherwise, were built during that period. When the light roadway of the Menai Suspension Bridge was blown down during a hurricane in 1839, it was essentially rebuilt as a matter of forced maintenance. The premier example of the state of the art of iron-chain suspension-bridge building in Britain was for some time to remain that of Telford's example across the Menai Strait. Iron-arch bridge concepts had evolved considerably in the half-century since Iron Bridge, and cast-iron girder designs had even been developed for more modest spans, like those leading up to the Dee Bridge.

In the mid-1840s, during the resurgence of railroad and bridge building known as railway mania, Robert Stephenson was called upon to design a number of bridges to meet a variety of water-crossing problems on the route of the Chester and Holyhead Railway, the critical rail link between London and the mail boats to Dublin. Among the bridges needed was the one to cross the River Dee at Chester. This bridge must have appeared to present structural problems almost trivial in comparison to the monumental problems that presented themselves to Stephenson in the design and construction of another bridge with which he was then engaged, a new crossing of the Menai Strait. The sudden collapse of the Dee Bridge in 1847 no doubt caused considerable concern that something was being overlooked also in the much more ambitious design, and this probably made the engineer more conservative in the large bridge then under construction than in his prior works.

The Britannia Bridge

In the early 1840s, amid the frantic activity associated with railway mania, the Chester and Holyhead Railway wished to have a permanent rail link across the Menai Strait so that ferries would not have to be employed on the line until the terminus at Holyhead, on the western side of the Isle of Anglesey. The existing Menai Suspension bridge had been carrying carriage traffic for about two decades, with only infrequent interruptions to rebuild its roadway after severe storms. Although these were minor inconveniences that closed the carriageway only until the light roadway could be rebuilt, similar interruptions of rail traffic would have been an embarrassment for a railway whose reputation depended upon reliability.

Furthermore, the great flexibility of the suspension bridge's roadway made it problematic for heavy locomotives to pull railroad trains across it. The weight of the locomotive itself would have deflected the bridge so much that a veritable valley would have formed under the load and the engine would have had to climb a steep hill to complete the passage. This was contrary to the objective of presenting a flat or at most a gradual grade to the steam locomotive. Stiffening the bridge apparently was not considered a viable solution; that would have added a great deal of structural dead weight that the chains were not designed to carry. George Stephenson, the great railway engineer and father of Robert, proposed unhitching the heavy locomotive when it reached the bridge, hitching up a team of horses to pull the cars across, and then connecting up another lo-

Figure 7.5. Telford's scheme to construct an iron-arch bridge without traditional falsework (Beckett, 1984; by permission of David & Charles, Newton Abbot, Devon, England).

comotive on the other side of the strait. This would have presented a resolution of the dilemma, but it would hardly have been in keeping with the image the railways were trying to project. What was desired was a bridge that was strong and stiff enough to enable heavy engines to pull fully loaded trains without interruption across the strategic body of water.

Because the Menai Strait was so important to the British Admiralty, it had considerable control over what kinds of obstructions could be placed in the path of tall-masted ships. For almost half a century engineers had been confident that iron-arch bridges with single spans of the order of 1,000 feet (and thus capable of spanning the strait without intermediate supports) were perfectly feasible, but the Admiralty did not wish to have the shipping channels obstructed by any falsework for what might be years of construction. As early as 1810 Thomas Telford devised a scheme whereby an iron arch could be constructed without anchoring falsework in the water by suspending the incomplete arch from the top until it was self-supporting (see Fig. 7.5). However, the Admiralty finally objected to an arch bridge regardless of its mode of construction, for the completed span would have reduced the headroom toward the shores, where the ships were sometimes driven by the tricky currents in the strait.

Stephenson was led to reconsider a suspension bridge, and in wondering if it were possible to stiffen the platform sufficiently for railroad trains he reflected upon some then-recent examples that

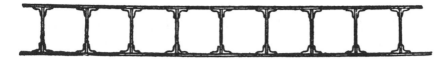

Figure 7.6. Cross section of Robert Stephenson's concept for a bridge over the River Lea at Ware (Clark, 1850).

employed trussing or trellising to carry turnpike roads and canals. Such schemes seemed to Stephenson to be difficult to extend to carry railroad traffic, however, for he believed that the extreme load on the suspension chain would alter its curvature so dramatically that (as quoted in Clark, 1850) "the direction and amount of the complicated strains throughout the trussing [would] become incalculable as far as all practical purposes are concerned." In other words, in the absence of analytical or computational tools to predict how close to failure the design would be, Stephenson did not feel comfortable proceeding with a suspension bridge.

The functional and political objections to existing bridge designs forced Stephenson to look to novel ways to span the Menai Strait. As designers are wont to do in the conceptual stages of design, Stephenson thought about the ways in which similar problems had been solved in the past, and he recalled a structural scheme he had employed in 1841 for a small bridge that spanned the River Lea at Ware. In order to meet the dual design constraints of maintaining sufficient headway above the towing path along the riverbank while not introducing a very great rise above street level, Stephenson had had to come up with a flat yet shallow bridge deck to span the fifty-foot distance. The scheme that he devised employed a wrought-iron platform, as shown in Figure 7.6, that he described as "a series of cells, . . . the whole being of boiler plate, riveted together with angle-iron, as in ordinary boiler building." Although the bridge actually built at Ware employed parallel girders in a more conventional way, the cellular design remained in Stephenson's mind. At first, he thought to adopt the cellular structure to provide a stiff platform for a suspension bridge, "and the first form of its application was simply to carry out the principle described in the wooden suspended structure, . . . substituting for the vertical wooden trellis trussing, and the top and bottom cross braces, wrought-iron plates riveted together with angle-iron."

Like most conceptual designs, that for the Britannia Bridge across the Menai Strait, once envisioned, could be sketched and articulated and criticized. Indeed, the design process then became one of successive identification of failure modes and modifications to obviate them. However, in the evolution of even the most radical of designs there is often an underlying constant: Robert Stephenson's unchanging idea soon became to span the distances (of the order of 500 feet) between towers to be erected on the shores and the Britannia Rock, which stood in the middle of the strait and thereby already obstructed shipping, with tubes of wrought iron that would be strong enough to support not only themselves but also the weight of railroad trains. These tubes would essentially be iron girders of such unprecedented size that the trains could pass through rather than go over the structure. Not only their form but also the manner of erection of the tubes would involve untried and thus unproven technology. According to Clark's (1850) account, even as the bridge was being completed it was

difficult to retrace chronologically the numberless modifications which the general outline of this curious structure had undergone; but whether an elliptical or rectangular section, or a combination of the two, was the subject of discussion, the tube was always treated as a beam, of which the wrought-iron top and bottom were analogous to the top and bottom flanges of a cast-iron T-girder. The calculations of its strength were made on this assumption.

Even the accounts of Clark and Fairbairn, voluminous as they are, only touch on the variations considered.

Whatever the chronology, as with virtually all innovative designs the form of the Britannia Bridge evolved by successive considerations of failure. With each conceptual iteration of the testing of a model, new modes of failure could be imagined possible and thus had to be checked against or obviated. Even after their form was fixed, the manner of getting the tubes in place remained an obvious problem. Furthermore, in the interests of getting the railroad link finished as early as possible, the project proceeded on the principle of what today is called "fast-track construction," and thus the stone towers were begun well before final details of the superstructure were known. An early concept was to employ chains like those used in suspension bridges to support the tubes during erection and provide supplementary support during use. When such chains proved to be unnecessary, the towers remained confusingly tall and have been found aesthetically objectionable by some structural critics (cf. Billington, 1983).

Because the use of tubular girders was an untried concept, Stephenson involved both Fairbairn and Hodgkinson, whose experiments and analytical interpretation of the experimental data, respectively, provided guidelines for design and extrapolation from scale models. At one stage Stephenson thought a rectangular form to be undesirable because it would need diagonal bracing that would not be consistent with leaving a clear passage for trains; furthermore, the large flat sides of the bridge would present a great exposed surface to the wind. Such concerns over features leading to failure led Stephenson to favor a circular or oval form, as suggested by nature and adopted in such artifacts as bamboo fishing poles. Early experiments were in fact carried out on circular tubes (see Fig. 7.7). What each experiment showed was that under a central transverse load applied to the bottom of the tube (simulating a heavy locomotive on a full-scale bridge), the tube failed by buckling at the top or by being torn asunder at the bottom. Such failures were exactly what the experiments were designed to reveal, for without knowing how and under what load the model tubes would fail, Stephenson and his fellow engineers did not know how to design the details of the bridge itself. As each model tube failed, the location of the failure was reinforced with additional thicknesses of wrought iron, and the ex-

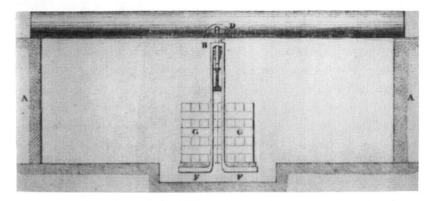

Figure 7.7. One of William Fairbairn's tests of model wrought-iron tubes for the design of the Britannia Bridge (Fairbairn, 1849).

periment repeated. It soon became evident that what was required to best obviate failure was a flangelike top and bottom, as used on cast-iron beams, and so the original rectangular form evolved back into consideration as more logical than the round.

In time, with progressively scaled-up experiments and Hodgkinson's resulting empirical formulas that predicted the failure load of a hollow rectangular beam, the bridge tubes themselves could be designed and assembled in the form shown in Figure 7.8, with flanges that showed some influence of Stephenson's abandoned Ware design. To design an experiment to test the full-scale tubes was essentially to build the bridge itself, and so, as with civil engineering structures generally, the final and crucial test would be the erection and proof testing of the unique structure. Because Stephenson and his colleagues focused so much on the limit loads of the bridge tubes, which created failure at the critical section at midspan, relatively little thought appears to have been given to the strength of the beam elsewhere along its length, although there was to be some debate between Stephenson and Fairbairn about how the tubes would behave over the piers. In the final design the bridge tubes were connected to form a continuous hollow girder, a feature the American bridge engineer Gustav Lindenthal (1922)

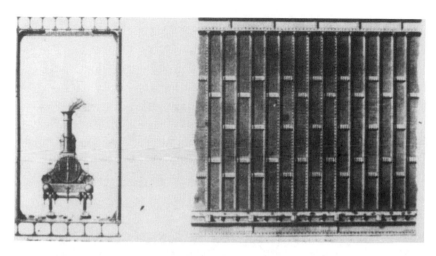

Figure 7.8. A tube of the Britannia Bridge: *left*, in cross section; *right*, side elevation showing arrangement of wrought-iron plates (Beckett, 1984; by permission of David & Charles, Newton Abbot, Devon, England).

later would state to have laid the foundation for the modern cantilever bridge.

Center-loading was always considered to be a worst-case situation, however, and it dominated the sizing of the tube, which resulted in a conservative design. Such conservatism naturally came at the expense of economy, and in the rush to design and build the Britannia Bridge questions of optimization of resources seem not to have been given as much thought as they might have. This occurred in spite of the fact that some iron rails had long been designed lighter at their ends than at midspan, thus economizing on material. In light of the fact that lifting the immense weight of the completed tube into place between the towers was to prove an enormous task, the tunnel vision that kept the tube almost constant in depth throughout its length proved to be a very costly error in design judgment. Furthermore, had the tube concept given way to one in which the train would ride atop a (box) girder, Stephenson's railroad bridge across the Menai Strait might have played an entirely different role in the history of structural design.

Figure 7.9. The Britannia Bridge, completed in 1850.

The Success and Failings of the Britannia Bridge

The great tubes of the Britannia Bridge showed themselves to be so robust during their floating and erection that supplementary suspension chains were never hung from the then superfluously tall towers (see Fig. 7.9). During proof tests, the bridge deflection was minimal, and the structural success of the tubular concept was total. Indeed, the bridge might still be carrying trains across the Menai Strait had not the heat of a fire in 1970 (started accidentally in a wooden roof added to protect the iron from water that would lead to rust) so deformed the wrought-iron tubes that it was deemed wiser to rebuild the bridge as an arch. While only a section of the original Britannia tubes was saved, to be put on display beside the rebuilt structure, the smaller but conceptually similar tubular bridge at Conway, Wales, erected on the same rail line and contemporaneously with the Britannia Bridge, remains standing to this day (see Fig. 7.10).

What made the bridge that stood for 120 years obsolete almost as

Figure 7.10. Stephenson's tubular bridge at Conway, Wales, with Telford's suspension bridge nearby (Pannell, 1977).

soon as it was completed was its designer's (and his collaborators') apparently consuming distraction, if not obsession, with the singular design problem of carrying trains across the Menai Strait in a wrought-iron tube. The enormity of this structural engineering problem in the climate of railway mania that existed when the project was conceived and approved seems to have so preoccupied everyone concerned that there was little time or inclination to ask questions relating to efficiency and use.

The total cost of the Britannia Bridge was approximately 600,000 pounds sterling in 1840s money. The experiments that played such

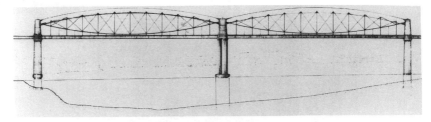

Figure 7.11. Isambard Kingdom Brunel's Saltash Bridge, also known as the Royal Albert Bridge, completed in 1859 (Beckett, 1984; by permission of David & Charles, Newton Abbot, Devon, England).

a critical role in making the bridge a structural success cost a mere 0.67 percent of the total. The material alone for the tubes consumed about 25 percent of the cost; their construction, which included the driving of about 2,000,000 rivets, consumed another 37 percent. When the cost of floating the tubes and raising them into place was added, the wrought-iron tubes of the Britannia Bridge accounted for almost 75 percent of the cost of the entire structure.

Even as the Britannia Bridge was being designed and constructed, alternative means of carrying railroad trains across 500-foot spans were under consideration by other engineers. Stephenson's contemporary, Isambard Kingdom Brunel, when faced with the problem of crossing the Tamar River at Saltash, in southwestern England, with a railroad bridge, came up with the solution of using large tubes as compression arches and wrought-iron chains as suspension chains in a self-equilibrating arrangement that, overall, looks somewhat like a lenticular truss (see Fig. 7.11). By the time Brunel's design was given the go-ahead, the free-spending days of railway mania had passed, and economy was at least as important as structure. Originally to be a two-track bridge, like Stephenson's across the Menai Strait, Brunel's bridge across the Tamar was finally opened in 1859 as a one-track span. The Saltash Bridge contained only 4,700 pounds of iron per foot of track length, compared to the Britannia's 7,000. In terms of pounds sterling

Figure 7.12. John Roebling's Niagara Gorge Suspension Bridge, completed in 1854.

per foot of single track, the cost of the Saltash was only about half that of the Britannia crossing (Shirley-Smith, 1976).

In North America, the engineer who had used the suspension bridge principle to carry canals over other bodies of water adopted the same principle to carry railroad trains across the Niagara Gorge (see Chapter 8). John Roebling's 820-foot span (Fig. 7.12) was opened to traffic in 1855 and provided a dramatic counterexample to the belief that a suspension bridge could not carry a railroad locomotive and heavy train effectively. Roebling went on to design and build a suspension bridge across the Ohio River at Cincinnati, and in his annual report for 1867 on that project he pointed out that whereas each 460-foot-long tube of the Britannia Bridge weighed over 1,500 tons, the entire 1,057-foot main span of his Ohio Bridge weighed less than 1,500 tons. Furthermore, the lighter suspension bridge was able to carry over 3,000 tons, which Roebling claimed was twice the live load that would "overtax the strength of a Britannia tube" (Roebling, 1867).

American and British bridge-building preferences were to diverge for the next century, with the British evolving variations on the truss-and-girder principle that were to lead both to milestone disasters like the collapse of the Tay Bridge, near Dundee, Scotland, in 1879 (see, e.g., Prebble, 1975; Sibly, 1977), and signal triumphs like the Firth of Forth Bridge, near Edinburgh, which was built strong in response to the fate of the Tay (cf. Koerte, 1992; Paxton, 1990). Americans, on the other hand, were to favor the suspension bridge, which evolved from the Niagara Gorge span into both the 1931 triumph of the George Washington Bridge over the Hudson River at New York and the 1940 embarrassment of the Tacoma Narrows Bridge. In all cases, economy was a principal determinant of form – until a colossal failure forced a reconsideration of the state of the art.

The concept of the Britannia Bridge proved to be a disappointment in more than economic terms. Because Stephenson's concept called for a tubular girder without penetrations, except at the extremities where railroad trains would enter and exit, each of the complete bridge's assembled tubes formed what was effectively a 1,500-foot-long wrought-iron tunnel supported 100 feet above the Menai Strait. This meant that the sun could heat the bridge's insides to uncomfortable temperatures, thus making the passage through it an exceptionally warm experience at best. Furthermore, the exhaust of the coal-burning steam engines filled the air with smoke and soot. Passengers did not find the passage through the Britannia tubes very pleasant at all, and the tubular bridge design was quickly identified as what might today be termed an environmental failure. One observer, writing in 1902 (and quoted in Beckett, 1984), described what he saw as he watched for a train to emerge from a wrought-iron tube:

Suddenly as one gazes, a hollow rumbling is heard, gradually increasing until with a hellish clang and the reverberation of a million echoes, a train dashes out, bringing with it a taste of the sooty air that lingers in the tubes, the product of fifty years, and abominably like that of an unswept chimney.

This shortcoming of Stephenson's design was incontrovertible, and there was a clear attempt to rectify it in one of the few other tubular bridges built after the Britannia. The Victoria Bridge was to cross the St. Lawrence River at Montreal, Canada, and it was to have about 6,000 feet of wrought-iron tubes as its major superstructure. Construction of the bridge was begun in 1854, and it opened five years later. One of the ways in which it differed from the Britannia Bridge was in having a slot formed in the top of the tube, as a means of dealing with the foul air. This did not prove to be very effective, however, and smoke was a "major nuisance," especially when combined with the temperatures of 125 degrees Fahrenheit recorded inside the tubes during the summer (Beckett, 1984).

Conclusion

All innovative designs can be expected to be somewhat uneconomical in the sense that they require a degree of research, development, demonstration, and conservatism that their technological descendants can take for granted. The case history of the Britannia Bridge provides a perspective on how the tunnel vision that can accompany a novel design project prevents designers from considering failure as carefully outside the narrow confines of the principal design challenge as they do inside it. This is not to say that there should be any less careful consideration of the major constraints of the problem; rather, it is to caution designers that they must make a special effort to step back from each design and consider what might appear to be some of the more mundane and less challenging aspects of the problem, those that might appear to lie on the periphery of the central focus. It would not appear to be inevitable that revolutionary designs need be doomed to economic or human-factors failure, for a circumspect design environment should properly consider those factors along with what appear to be the principal ones (cf. Norman, 1988).

Of course, many design climates are subject to the railway mania atmosphere, in which getting the job done within tight time constraints precludes the exploration or optimization of a design so as to obviate its obsolescence upon completion. A Chester and Holyhead Railway may have been willing to accept an overly heavy (and hence overly expensive) bridge in the interests of getting a jump on the competition or merely in the interests of getting the bridge in place so that it could contribute to the revenue base. But it seems unlikely that the railway would have chosen a hot and smoky tube over an open-air crossing as the bridge at Saltash proved to be. Indeed, in all the correspondence between Stephenson and Fairbairn documenting the testing of tube form after tube form and considering the structural principles of the Britannia Bridge, there appears to have been no consideration given to how it would function with people riding in the trains.

The value of a case history like this lies in its potential for contributing to a common heritage of design paradigms whereby the designers of today and tomorrow can gain a circumspection that might otherwise be learned only by repeating the mistakes of the past. A design problem like bridging the Menai Strait will never again occur in exactly the same technological, economic, and political climate as Robert Stephenson and his colleagues found themselves in a century and a half ago. However, the *process* of design, from the conceptualization to the realization of the artifact, is essentially a timeless, placeless, and even thingless human activity that succeeds only insofar as it anticipates failure. Distilling the essence of that process (and especially its pitfalls) from a classic case history like that of the Britannia Bridge can provide an appreciation for the limitations of design that is unlikely to come in any other way, except in repeating the mistakes of earlier designers. By understanding the kinds of errors and lapses in judgment that have limited the achievement of some of the great designers of the past, we are less likely to repeat those same kinds of errors in the future.

8

Failure as a Source of Engineering Judgment

John Roebling as a Paradigmatic Designer

The first and most indispensable design tool is judgment. It is engineering and design judgment that not only gets projects started in the right direction but also keeps a critical eye on their progress and execution. Engineering judgment, by whatever name it may be called, is what from the very beginning of a conceptual design identifies the key elements that go to make up an analytical or experimental model for exploration and development. It is judgment that separates the significant from the insignificant details, and it is judgment that catches analysis going astray. Engineering judgment is the quality factor among those countless quantities that have come to dominate design in our postcomputer age. Judgment tells the designer what to check on the back of an envelope and what to measure at the construction site. Judgment, in short, is what avoids mistakes, what catches errors, what detects flaws, and what anticipates and obviates failure. The single most important source of judgment lies in learning from one's mistakes and those of others.

All meaningful improvements in analytical and computational capabilities are at heart improvements in our ability to anticipate and predict failure. Every engineering calculation is really a failure calculation, for a calculated quantity has meaning for engineering only when it is compared with a value representing a design constraint or failure criterion of some kind. Indeed, a factor of safety can only be calculated within the explicit context of a failure mode, and the most successful designs are those that involve the most complete proactive failure analysis on the drawing board in the design office. The surest way of conducting as complete as possible a proactive failure analysis is for those individuals most involved with a design to have as broad and sound a range of engineering judgment as the project demands.

Engineering judgment does not necessarily come from a deeper understanding of theory or a more powerful command of computational tools. The traditional engineering sciences are servants and students but not masters and teachers of engineering judgment, and even design experiences in an academic context can provide but limited perspectives on the meaning of judgment. The single most fruitful source of lessons in engineering judgment exists in the case histories of failures, which point incontrovertibly to examples of bad judgment and therefore provide guideposts for negotiating around the pitfalls in the design process itself. Another invaluable source of lessons in sound engineering judgment comes from the great engineers, who by their works have demonstrated that they possessed impeccable judgment, which has more often than not come from their critical study of failures and near-failures.

Engineers have always had to wrestle with matters of knowledge and ignorance, and the intellectual problems of design were no less complex in the past than they are today. Indeed, it might be argued that in an environment of lesser analytical understanding and lesser computational potential, the intellectual challenge of design was even greater than it is today. Some of the engineers of the past whose achievements we admire have left not only the artifacts they designed but also an intellectual record of how they designed them. In such records are the lessons of the masters, and they are no less relevant today than they were when they were written. Thus it behooves us to look into these records to distill at least their potential for conveying a sense of engineering judgment. When this is done it becomes clear in case after case that the proper anticipation of failure has always been the mark of the most successful of engineers, and that lack of judgment is what leads to failures.

In this chapter the example of the great nineteenth-century bridge engineer John A. Roebling will be considered. He was not only prolific as a bridge designer and builder but also as an author of reports on his design activity. His writing shows him also to have been an avid reader of and listener to accounts of failures, as have been great engineers generally. They have recognized that it is possible to learn from the mistakes of others what constitutes lack of judgment and thus by inference to increase their own judgment. No matter how long ago they lived or how long ago their mistakes were made, the engineers of the past remain the master teachers of judgment today (cf. Billington, 1977; Peck, 1969, 1981).

John Roebling as a Paradigmatic Designer

The story of John Roebling, and his son and daughter-in-law who saw the Brooklyn Bridge through to completion, has been told admirably by the bridge engineer David Steinman (1950) and by the writer David McCullough (1972). Such full-length biographical studies are essential for gaining an in-depth appreciation of the intellectual growth and development of an engineer, and they are recommended reading for anyone seeking a fuller appreciation of the engineering character and the development of judgment. Such general works can also provide an appreciation of the design challenges and the technical (and nontechnical) obstacles that had to be overcome in bringing to fruition what were only the dreams of earlier engineers. But these general works necessarily stress the human and social aspects of the engineers and engineering. To gain the greatest insight into how some of the most successful engineers thought about the design and execution of their greatest achievements, it is necessary to go to the writings of the engineers themselves.

John Roebling (Fig. 8.1) was born in 1806, in the German region of Saxony, and was educated at the Polytechnic Institute in Berlin. There he studied hydraulics and bridge construction, among other subjects, including philosophy under Hegel, who reportedly thought highly of Roebling as a student and first got him to think about going to America. Roebling did come to America, in 1831, and settled in western Pennsylvania. At first the town he founded was called Germania, but later its name was changed to Saxonburg, and Roebling believed it could be built into an "earthly paradise." His destiny was not in farming, however, and eventually he started a factory to manufacture wire rope, which would be stronger and more reliable than the hemp rope that he noted was failing and causing many deaths on the canals that were such important transportation routes before the railroads displaced them. In 1849, the thriving wire rope business was relocated to Trenton, New Jersey, from where it would play a major role in the history of bridge building in North America (cf. Steinman, 1950; McCullough, 1972).

Roebling was an especially articulate and prolific writer, although his audience was more often the boards of directors of the projects he was engaged in than the engineering profession or the general public. To see the million-dollar projects that he conceived actually become fully operating structures, he necessarily had to interact with politicians and businessmen and convince them that what was proposed was not only structurally sound but also politically and financially sound. Roebling seems to have been especially successful in communicating his sense of good judgment to nonengineers because he was so explicit and clear about the method by which he reached his design conclusions.

Like bridge builders before and after him, Roebling had to answer tough questions about success and failure before any rights were granted or any investments made. For all his early reputation as a

Figure 8.1. John A. Roebling (Steinman, 1950).

successful builder of suspension bridges to carry canals over natural waterways, Roebling had to argue anew his case when in the mid-nineteenth century he proposed an 800-foot suspension bridge to carry not the uniform and calm load of water but the irregular and violent load of steam locomotives and railroad cars over the Niagara Gorge (see Fig. 7.12), or a 1,000-foot bridge to span the Ohio River at Cincinnati and leave no obstructions to the growing barge traffic, or a 1,500-foot bridge across the busy East River that would eliminate the need for ferries, which were especially unreliable during winter months, between New York and Brooklyn.

Before Roebling could approach politicians and financiers, however, he had to have the technical confidence that a suspension bridge was indeed a viable design alternative. After all, in the 1840s the dependability of suspension bridges was seriously questioned by engineers and lay persons alike. Even the most graceful and tech-

nologically advanced wrought-iron chain suspension bridge built by
Thomas Telford across the Menai Strait in northwestern Wales (see
Fig. 7.4) had experienced trouble in high winds, with its deck undergoing severe vibrations shortly after it opened in 1826. Ten years
later the deck was observed to oscillate with a double amplitude of
sixteen feet during a storm, and in 1839 a hurricane brought down
part of the roadway.

A recreational chain pier, which was really a suspension bridge
that jutted out from the beach into the sea at the resort town of
Brighton in southern England (Fig. 8.2), had experienced similar
damage in an 1833 gale (Fig. 8.3). Its deck was also damaged in an
1836 storm, and the destruction was described and illustrated (Fig.
8.4) in a paper by J. Scott Russell (1839), who theorized that the
failure was caused by excessive vibrations. Other horror stories of
suspension bridges failing under the marching feet of soldiers, the
sudden movement of people watching boat races, or the stampede
of cattle appeared regularly in the pages of newspapers, magazines,
and technical journals.

In the wake of such disasters, the suspension bridge was naturally
considered with some skepticism, and it came to be regarded with
such disrepute in Britain that elaborate alternative designs like the
Britannia tubular bridge were undertaken at great expense. Robert
Stephenson's concept of great wrought-iron tubes spanning 500-foot
distances actually grew out of a suspension-bridge design of Roebling's for a suspended canal crossing (see, e.g, Clark, 1850), but the
fear of failure associated with the lighter bridges led Stephenson to
develop his tube concept into a self-supporting hollow girder that
set the direction in which British bridge building would go for the
next century. With some notable exceptions, such as Brunel's classic
design for a bridge across the Avon gorge (see Fig. 8.5), which like
Telford's across the Menai Strait was to carry only carriage traffic,
the suspension bridge was to be excluded from consideration by the
conventional technical wisdom in Britain whenever a stiff and reliable bridge for railroad traffic was needed.

Figure 8.2. Brighton Chain Pier (Bishop, 1897).

Roebling, however, did not accept the British conventional wisdom. Although it is generally agreed that the principle of the suspension bridge may date from early times, when vines and other
organic cables were anchored to trees, the modern impetus in North
America is attributed to the engineer James Finley, who in the early
1800s suspended stiffened-truss decks from wrought-iron chains.
Whether it was the American roots of the modern design or the fact
that Roebling was in the business of making wire rope, which would
replace the wrought-iron chains or at least provide support and bracing for the roadways of some of his early suspension bridges, in the
1840s he refuted the British claims and was confident that substantial
suspension bridges could be erected to carry railroad trains. In a
letter to a bridge company formed to erect a bridge across the Niagara Gorge, he wrote (as quoted in Pugsley, 1968):

Although the question of applying the principle of suspension to railroad
bridges has been disposed of in the negative by Mr. Robert Stephenson
. . . any span with fifteen hundred feet, with the usual deflection, can be made
perfectly safe for the support of railroad trains as well as common travel.

Figure 8.3. Brighton Chair Pier after 1833 gale (Bishop, 1897).

Roebling could make such assertions not only because of his already extensive experience with suspended canal bridges around Pittsburgh and elsewhere but also because he had studied the nature and causes of the failures of the suspension bridges that had cast doubt on the form. Rather than take the failures as mere evidence that the suspension bridge principle was ill suited to carrying railroad trains, Roebling looked at the failures as found or fortuitous experiments that revealed what most needed attention in designing against failure (see Roebling, 1841).

Even though there were many suspension bridge failures in the first part of the nineteenth century, each new one was an event of interest to engineers. In a paper published in the *Transactions of the Institution of Civil Engineers* for 1842, C. W. Pasley, a colonel in the Royal Engineers, describes traveling in Scotland during a terrible hurricane and hearing that the suspension bridge at Montrose had been destroyed by the storm. He altered his route to observe the damage:

SKETCH Shewing the manner in which the 3rd Span of the CHAIN PIER at BRIGHTON undulated just before it gave way in a storm on the 29th of November 1836.

SKETCH Shewing the appearance of the 3rd Span after it gave way.

Figure 8.4. Brighton Chain Pier during and after 1836 storm (Russell, 1839).

I was induced to stop there, that I might have an opportunity of inspecting the construction of that bridge, and of ascertaining from what cause, or, at least, in what part, it had given way; having always been of the opinion that from the examples of failures some of the most instructive lessons in practical architecture or engineering are to be derived.

Pasley goes on to relate the circumstances surrounding an earlier failure of the bridge at Montrose. Shortly after it opened, one of its chains gave way when a crowd of spectators watching a boat race rushed from one side to the other of the bridge. In response to that failure, it "having proved the insufficiency of the chains in their existing state," the suspension chains had been strengthened. Pasley also discusses a distinction between the terms "vibration" and "undulation," both of which were being used, sometimes interchange-

Figure 8.5. Isambard Kingdom Brunel's Clifton Suspension Bridge over the Avon gorge, realized posthumously in 1864.

ably, to describe the behavior of suspension bridges in storms. He also discusses the benefits of stiffness against vertical motion of bridge decks, and contrasts the construction of failed and successful bridges, noting deficiencies of the former when compared with the latter. In a subsequent communication, J. M. Rendel (1841) also discusses the failure of the Montrose bridge and argues for "the importance of giving to the roadways of suspension bridges the greatest possible amount of stiffness" to prevent failure during violent storms.

Pasley also comments on Telford's Menai suspension bridge, whose damage in an 1839 storm was described by another member of the Institution of Civil Engineers, W. A. Provis. From that (1842) account Pasley concludes that the Menai bridge did not prove to be as effective in the wind as was expected. But Pasley is quick to point out that by criticizing the design of the bridge, he meant "no disparagement" to his friend Telford, the first president of the Institution, for when his bridge was designed and constructed engineers possessed "no experience of a suspension bridge of so very large a span." Furthermore, "in construction entirely and absolutely new, no man, however great his talents may be, can be expected to foresee every contingency that ought to be provided for." Pasley was conceding, in other words, that failures could be expected to happen, especially when designs went beyond the state of the art. In a note appended to his paper, Pasley reports being informed that in the design stage Telford did consider "trussing the roadway" of his bridge but "finally decid[ed] upon omitting it in the first instance, and adopting it subsequently should experience prove it to be necessary." Such design modification in response to failure is, of course, much more common in mass-produced items than in large and unique civil engineering structures like suspension bridges (cf. Petroski, 1992f). In the latter, it is clearly preferable from the outset to err on the side of conservatism, although examples of retrofitted bridges do exist (cf. Figs. 9.5 and 9.6).

Roebling, who in the 1840s was engaged in making wire rope, published a paper of his own in the *American Railroad Journal and Mechanics' Magazine* (Roebling, 1841). His "Remarks on Suspension Bridges, and on the Comparative Merits of Cable and Chain Bridges" begin with his asserting that "storms are unquestionably the greatest enemies of suspension bridges," and continue with his pointing out that there was then a "diversity of opinion among Engineers" as to the effects of the wind on the structures. Roebling criticizes the flexibility of chain bridges and notes that "the floors of almost all the English suspension bridges are entirely too light." He indicates the weaknesses of suspension bridge designs, including that of the Menai, and describes some ways in which cables, as opposed to chains, could be employed to steady a bridge against the wind and unwanted vibration. In closing his paper, Roebling explains somewhat apologetically why he is pointing out so many negative qualities of suspension bridges as then built:

Figure 8.6. One of Marc Brunel's suspension bridges (Hopkins, 1970).

The above remarks have not been made with a view of bringing suspension bridges into discredit. To impute such a motive to me would be unjust. No one can be a greater admirer of the system than myself. . . .

In speaking of the weak points of the system, I have only intended to show how much caution is necessary in planning and executing a suspension bridge in order to insure perfect safety.

Roebling recognized that to build a successful bridge he had to anticipate how it could fail. By being completely familiar not only with the events but also the causes of the problems with past suspension bridges, an engineer like Roebling was able, in spite of Pasley's skepticism, to anticipate as much as humanly possible what could go wrong with a suspension bridge and incorporate features in the design that would obviate modes of failure. He was using failure analysis in its most constructive and proactive form.

It was not only failures that attracted the attention of Roebling, of course, for no engineer can be so obsessed with failure or petrified by its possibility that successful designs go unnoticed or unbuilt. Among the bridges Roebling admired were those that Marc Isambard Brunel, Isambard Kingdom Brunel's father, designed earlier in the century on the Isle of Bourbon, in the Indian Ocean, where hurricanes were common. These two bridges, which Roebling read about in Navier's 1823 memoir on suspension bridges (cf. Picon, 1988), were of modest size, one with two spans of 122 feet each and the other with a single span of 131 feet (Fig. 8.6). Above the roadway they were of more or less conventional construction, with the suspension chains arranged in groups to allow the roadways to be suspended. The most distinguishing feature of the elder Brunel's bridges was underneath, however, where the roadway was connected to "reversely-curved chains, the ends of which are secured in the piers and abutments. At the same time the ends of these reversed chains recede from the bridge horizontally, and act as diagonal stays in the latter direction" (quoted in Roebling, 1841). In other words, the bridge deck was prevented from moving vertically or horizontally in the wind by the restraint that was applied from below and from the sides.

The idea of tying down the deck of a suspension bridge to suppress undesirable movement or vibrations was also the subject of the paper read before the Scottish Society of Arts in 1839 by John Scott Russell (Fig. 8.7). In this paper, Russell describes how a violent storm had caused "considerable damage from the vibrations produced by the action of the wind" to a timber scaffolding made up of long, slender members, which led him to look into the causes of the failure. Subsequent experimental observations on such a structure revealed to Russell that it vibrated in distinct modes that could be changed and mitigated by altering the positions of the crossbars. He draws analogies with the vibration of elastic cords and strings in the wind and the effects of a finger on them, concluding that if the vibrating body were constrained at any place but a node its motion could be considerably restricted. He applies his observations to suspension bridges and considers various cases of arrangements of stays.

Whether or not Russell's paper was read by Roebling, he came up with a scheme for a bridge over the Niagara Gorge that employed radiating wire stays that were anchored in the rock below the two-tiered roadway (see Fig. 7.12). Both the stays and the deep and stiffening deck structure can be seen in a contemporary photograph (Fig. 8.8), and the bridge proved to be a signal success. Indeed, the fact that the Niagara Gorge bridge deflected only three and one-half inches under the first locomotive (weighing twenty-three tons) that passed over it was an incontrovertible counterexample to the British hypothesis that a suspension bridge could not carry a railroad train.

Fig. 11.

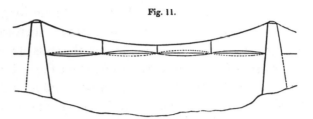

Now, in order to prevent these oscillations, it will be of no use to adopt the various methods of staying that have hitherto been adopted ; it is of no use to carry a stay-chain to the middle of the roadway, fig. 12, nor even four stays as at 2, 3, and 4,

Fig. 12.

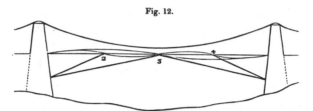

because it will vibrate exactly as at figs. 9 and 11, neither would stays at 2 and 3, as in fig. 13, produce the effect, because the

Fig. 13.

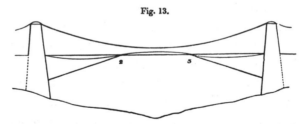

bridge would still oscillate as in fig. 10.

Mr Brunel has proposed a method of preventing oscillation,

Figure 8.7. A page from John Scott Russell's paper illustrating how a suspension bridge can vibrate in the wind and how cable stays are not fully effective if located at nodes (Russell, 1839).

Figure 8.8. Contemporary photograph of the Niagara Gorge Suspension Bridge showing the stiffened roadway and the anchoring stays.

Roebling had achieved with uncommon success what other engineers had agreed could not be done. His principal advantage was not a more sophisticated theory or more careful calculations. Roebling achieved his success by concentrating his design judgment on how his bridge might fail. He kept the deck from deflecting by making it stiff; he kept the roadway from being destroyed in the wind by making it heavier than the Menai deck and by tying it down. In short, Roebling achieved success by identifying and confronting failure modes as the principal part of his design charge. These were not merely design-theoretical speculations after the fact, for in his own writings Roebling himself provided ample evidence that it was through studying failures that he achieved success.

After the principles embodied in the Niagara Bridge had been established as sound, Roebling (1855) wrote, in a final report that Steinman (1950) called "a classic in engineering literature":

Professional and public opinion having been adverse to Suspended Railway Bridges, the question now turns up, what means have been used in the Niagara Bridge, to make it answer for Railway traffic? The means employed are *Weight, Girders, Trusses,* and *Stays.* With these any degree of stiffness can be insured, to resist either the action of trains, or the violence of storms, or even hurricanes; and in any locality, no matter whether there is a chance of applying stays from below or not. And I will here observe, that no Suspension Bridge is safe without some of these appliances. The catalogue of disastrous failures is now large enough to warn against light fabrics, suspended to be blown down, as it were, in defiance of the elements. A number of such fairy creations are still hovering about the country, only waiting for a rough blow to be demolished.

Roebling goes on to state that, by using the best quality of steel wire, suspension bridges of almost a mile span could be constructed and "offer the same degree of security" as the Niagara Bridge. His own Brooklyn Bridge, completed posthumously almost thirty years later, spanned over a quarter of a mile between towers, and within another century the mile span was being approached in structures like the Humber Bridge, completed in 1981 in eastern England. More than 4,600 feet between towers, the Humber will remain the longest suspended span in the world until the Japanese complete construction on a mile-long span toward the end of the century. In the meantime, of course, such collapses as that of the Tacoma Narrows Bridge occurred, in large part because bridge designers had ignored or forgotten Roebling's observation that no suspension bridge was safe without some of the "appliances" that he learned to add in order to obviate the failures he had studied in the "catalogue of disasters."

Further evidence of Roebling's conscious concern with failure as a guide to success emerged in the wake of the destruction of the 1,010-foot main span of the Wheeling Suspension Bridge, designed by Charles Ellet, Jr. In 1847 both Roebling and Ellet had submitted bids for the construction of a bridge to cross the Ohio River at Wheeling. Ellet's light and narrow 1,010-foot-long design was accepted over Roebling's more conservative proposal, and when the bridge was completed in 1849 it was the longest span in the world. Five years later, in May 1854, the deck of the Wheeling Bridge was tossed about in a tremendous storm until the structure collapsed.

In the meantime, Roebling had won the contract for the Niagara Bridge, and it was under construction when the Wheeling failure occurred. Naturally there were concerns that the same fate lay in store for the new bridge, and Roebling addressed them head on in a section on effects of high winds in his report of 1855, shortly after the bridge was officially opened to traffic:

It is my duty to establish the safety of the Niagara Bridge, which has already, and in the brief space of one month, become one of the greatest thoroughfares on this continent. I cannot do so without drawing a comparison with other works, and without pointing out the defects which caused the destruction of the Wheeling bridge, and on the other hand explaining the means of safety, which have been employed in the Niagara Bridge.

[T]here was no provision in the whole structure [of the Wheeling Bridge] aside from the inherent [but inadequate] stiffness of the floor, which could have had an effect in checking vibrations. . . . The destruction of that bridge was clearly owing to a want of stability, and not to a want of strength. This want of stiffness could have been supplied by over-floor stays, truss railings, under-floor stays, or cable-stays. If by these means no high degree of stiffness could have been obtained, they would at any rate have proved quite sufficient to check oscillations, and to keep them within safe limits. In the Niagara Bridge most ample provisions for stability have been made.

Roebling clearly establishes the soundness of his design by contrasting its features with those that he found wanting in a bridge that failed. This kind of recitation of how potential failure modes have been obviated is indeed the only logical (as opposed to physical) way to "prove" that a design is safe.

Any argument for the safety of a design that relies solely on pointing to what has worked successfully in the past is logically flawed.

For example, until it was destroyed in the storm, the Wheeling Bridge could have been considered a successful structure and could have been acclaimed as a model for successful bridge design. But any suspension bridges that were mere copies of the Wheeling, supposedly incorporating its successful features, would have in fact repeated the latent weaknesses and flaws of its design. No matter that the Wheeling Bridge stood for five years, for in those five years it was simply not tested to the extent that it was in the storm of 1854. Before May of that year, any and all assertions of a Roebling that the bridge was unsafe could have been dismissed by the observation that the bridge was still standing and that his warnings were sour grapes. It was only in the failure of the bridge that the flaws in its design were rendered patently obvious. Only the failure provided an incontrovertible counterexample to the putative hypothesis that the Wheeling Bridge was a successful design.

Roebling's repeated successes in suspension bridge design can all be attributed not to his copying his own safe designs, for he simply did not do that, but to his constant and explicit concern with the avoidance of failure as a guiding principle in design. In a report submitted September 1, 1867, to the president and directors of the New York Bridge Company, Roebling begins his discussion of the practicability and strength of the bridge he proposed to link New York and Brooklyn by describing the breaking strength of "a bar of good wrought iron." After establishing that wire drawn from such a bar could support, at its breaking strain, 32,400 feet of its own length hung vertically, he applies a factor of safety of three and argues that one-third of that length, or 10,800 feet, "will, if left undisturbed and kept from oxidation, support its own weight any length of time." When stretched "across a wide chasm," the increased tension in the wire must be taken into account, of course, as Roebling argues in the posthumously published report (1870):

The cables of the East River Bridge will have a deflection of 128 feet, which is $\frac{1}{12.5}$ of the span, and the tension which is produced thereby, will be about $1\frac{2}{3}$ of the weight of the wire. In order now to ascertain the length of span which may safely be attempted to a deflection of $\frac{1}{12.5}$, we have to reduce the above 10,800 feet to 6,480 feet. In other words, the best quality of iron wire, if suspended over a chasm of about 6,000 feet, and with a deflection of $\frac{1}{12.5}$ or 480 feet, will only be exposed to one-third of its breaking strain. . . .

From the above simple facts and considerations, it is plain, that the central span of the East River Bridge, which is only 1,600 feet from centre to centre of tower, is far within the safe limits of good wire. A span of 1,600 feet or more can be made just as safe, and as strong in proportion as a span of 100 feet.

Roebling's judgment clearly had a strong sense of conservatism, based explicitly on the concepts of failure limits and margins of safety, but this in no way kept him from proposing bold new designs. John Roebling's clear sense that it was always failure that he was designing against was passed on to his son, Washington. Evidence of this can be found throughout the younger Roebling's 1877 report on progress of the Brooklyn Bridge. For example, when discussing the design of the anchor plates, the chief engineer explains why the weight of the masonry acting on the plates against the upward pull of the chains had the relatively low factor of safety of two and one-half:

This figure may appear small when compared with the main cables, where the margin of safety is *six* times, were it not that the conditions are essentially different. In the case of the cables, we have to provide for numerous contingencies which cannot occur in the anchorage. For instance, allowance must be made for the deterioration of the wire by the elements, for any possible imperfection in the manufacture of the cables, for any increase in weight or strain over that first contemplated, which is most always the case, and lastly, a certain margin is required so as not to strain the wire beyond its limit of elasticity.

In the anchorage, however, we have only two factors to deal with – granite and gravity. The first, a material whose very existence is a defiance to the "gnawing tooth of time;" the second, the only immutable law in nature; hence, when I place a certain amount of dead weight, in the shape of granite, on the anchor plates, I know it will remain there beyond all contingencies.

A telling photograph, which was taken shortly after the catwalks were erected to the newly completed Brooklyn Bridge towers, provides further evidence that the son worried as much about failure as his father. These catwalks, which were necessary for conducting the cable-spinning operations, were a kind of suspension bridge in their own right. In the foreground of the photo (Fig. 8.9) is a sign that reads:

SAFE FOR ONLY 25 MEN AT ONE
TIME. DO NOT WALK CLOSE TOGETHER,
NOR RUN, JUMP, OR TROT, BREAK
STEP!
W. A. Roebling, *Eng'r in Chief.*

Although the elder Roebling died in 1869, of complications following an accident that occurred when surveys were being made for the Brooklyn Bridge, the preemption of failure embodied in the son's warning sign is a clear echo of the father's judgment. Both engineers, father and son, knew that concentrated loads would cause excessive deflections of an unstiffened catwalk, and that running, jumping, or trotting could send the long, light structure into uncontrollable vibrations that could throw the men off and tear the walk apart. In all his suspension bridges, John Roebling knew the value of a truss for stiffness and the importance of ties for suppressing vibrations. Since the catwalk could not practically have a truss, Washington Roebling provided the next-best thing – a warning to obviate what destructive forces he could. (The catwalk was expected to be destroyed by the uncontrollable force of very strong winds, and this was accepted as a matter of maintenance.) These are in fact the objects of all design: either to avoid destructive forces or to provide,

Figure 8.9. The Brooklyn Bridge under construction, with Washington Roebling's warning sign in the foreground (New-York Historical Society).

within understood limits, sufficient resistance to them in the structure. The actions of the wind on a full-scale bridge clearly could not be controlled, but the actions of men on a catwalk could. Because John Roebling understood what destructive forces the wind could exert on a full-scale bridge, and because he judged their effects to be dominant, he could and did design successful bridges by building into them the resistance to failure by the wind and other design loads. The Brooklyn Bridge (Fig. 8.10) stands today as a monument to engineering judgment.

Figure 8.10. The Brooklyn Bridge, completed in 1883.

Conclusion

Case histories of failures and strategies for failure avoidance provide an invaluable source of information about design that has generally not been exploited in more than an ad hoc way. John Roebling, who used lessons learned from failures to achieve success in his own designs, provides a valuable model of how a designer proceeds rationally. Not all engineers can be so fortunate as Washington Roebling in having a professional mentor as a parent, but we can all learn a sense of good design and judgment from the legacy that masters like John Roebling have left for us. When those engineers were as prolific and straightforward in their writing of reports as they were in their design of structures, they have left us a legacy in words as enduring as the granite in their bridges.

Because design is a process involving the anticipation and obviation of failure, the more knowledgeable designers are about failures the more reliable should be their designs. Since the laws of nature and physics are assumed to be immutable, any lessons learned from the behavior and especially the failure of even ancient designs are no less relevant today, and the good design practice of engineers in centuries past can serve as models for the most sophisticated designs of the modern age. Indeed, ignoring wholesale the lessons and practices of the past threatens the continuity of engineering and design judgment that appears to be among the surest safeguards against recurrent failures.

9

The Design Climate for the Tacoma Narrows Bridge

A Paradigm of the Selective Use of History

The design prejudices prevailing in the 1920s and 1930s, when the George Washington, Golden Gate, Tacoma Narrows, and other major suspension bridges were on the drawing board, evinced confidence in analytical techniques and a preoccupation with aesthetics. Almost a half-century of relatively successful experience with suspension bridges like the Brooklyn had led designers to focus on extrapolating from that successful experience. Such a focus led in turn to a climate in which some designers forgot or ignored the first principles on which the successful designs originally were based. These principles rested upon criteria that explicitly recognized wind-induced failure modes that in turn drove design decisions. Basing structural extrapolation upon models of success rather than on failure avoidance was to result in such flexible bridges as the Bronx-Whitestone, completed in 1939 over the East River in New York City, and culminate in the collapse of the Tacoma Narrows Bridge in 1940.

It is perhaps a truism that the design climate in which an engineering project is conceived and developed can have a profound effect on whether it succeeds or fails, but the significance of the interaction between success and failure in design can sometimes be so subtle as to escape even the greatest of designers. The context in which suspension bridges were being designed and built in the late 1920s and 1930s in the United States provides a classic example of the phenomenon.

The modern history of suspension bridges begins in the early nineteenth century and is embodied in such masterpieces as Thomas Telford's Menai Strait Suspension Bridge (Fig. 7.4), completed in 1825. For all the beauty and technological achievement embodied in this structure's 580-foot span hung from wrought-iron chains, it was plagued by an extremely flexible deck that tended to sway dangerously in high winds. As we have seen, such unwelcome behavior in the Menai and contemporaneous suspension bridges brought a bad reputation to the form, and when large spans were required to carry railways across the likes of the Menai Strait, revolutionary new forms like Robert Stephenson's tubular Britannia Bridge were devised in Britain.

In North America, Stephenson's contemporary John Roebling was not so ready to dismiss the suspension principle for carrying heavy railroad trains. As discussed in Chapter 8, by the mid-1850s he had designed and built the Niagara George Suspension Bridge, whose 820-foot suspended span not only proved to be stiff enough for railway traffic but also proved to be able to withstand high winds without damage. Unlike Stephenson and other British engineers, who took the weaknesses and failures of suspension bridges as indications that the form was unsuited to the function, Roebling studied the faults and flaws of earlier bridges to learn what it was he had to avoid in his own designs. His suspension bridges worked wonderfully, of course, culminating in his masterpiece, the Brooklyn Bridge, in 1883.

The success of Roebling's bridges inspired many another engineer to propose still longer suspension spans, but these designers appear to have focused on the success of designs like Roebling's rather than on the fundamental role that proactive failure analysis played in his design thinking. Thus, in a matter of a generation or two of major bridge designers, the design climate changed from one in which the occurrence of failures had to be explicitly addressed by anyone proposing a new suspension bridge to one of such confidence and even hubris that designers took off from the achievements of Roebling and his successors in directions that appear almost at times to have focused more on economy and aesthetics than on structural behavior.

The design climate in the 1920s was epitomized in the George Washington Bridge project, and was carried to even further extremes later in such suspension bridges as the Bronx-Whitestone, Deer Isle, and Golden Gate, all completed in the late 1930s. The limits of ignorance were of course reached in the Tacoma Narrows Bridge, whose misbehavior in the wind and ultimate collapse only months after opening in July 1940 constitute one of the most tragic episodes in all of engineering history.

Billington (1977) has treated this subject in considerable depth, and his paper generated much discussion in the *Journal of the Structural Division* of the American Society of Civil Engineers. Among the discussion participants were some of the most prominent bridge engineers of the mid-twentieth century, several of whom had worked with Othmar Ammann, the engineer who designed the George Washington Bridge, which epitomized the state of the art. They took offense at Billington's criticism of the individual engineer Ammann as opposed to the entire contemporary profession, but Billington made a strong case in his conclusion for Ammann's prototypicality in particular and the role of structural criticism in general. He also argued eloquently for the importance of the history of engineering in the development of design and the design climate. His concluding observations that "individual personalities play a significant role in the history of structures, even though such individuals are partially constrained by the context in which they work," and that "history, for structural engineers, is of importance equal to science," might well serve as epigraphs for this chapter and, indeed, this entire book. The reader is encouraged to see Billington (1977) for elaboration.

Design Climate for a Hudson River Crossing

Othmar Ammann, chief bridge engineer of the Port of New York Authority, which built, owned, and operated the George Washington Bridge across the Hudson River, himself describes in some detail the general conception and development of the bridge's design in an article that appeared in the *Transactions of the American Society of Civil Engineers* in 1933. He begins by relating the history, legislation, and financing of the project that culminated in a spectacular bridge with a suspended span of 3,500 feet, which at the time was twice as great as that previously achieved, in the 1,595-foot Brooklyn Bridge a half-century earlier (see Fig. 3.3) and its neighbors, the Manhattan Bridge, completed in 1909 with a span of 1,480 feet, and the Williamsburg Bridge, completed in 1903 with a record 1,600-foot suspended span. Ammann was to prove himself a master of the genre, and his description of suspension bridges generally and the conception and development of the George Washington Bridge in particular provides a snapshot of the design climate of his times.

Whereas Roebling's reports in the middle decades of the nineteenth century frequently referred to the oscillation and collapse of famous suspension bridges, Ammann's report avoids explicit mention of structural design deficiencies, flaws, weaknesses, and failures of bridges of any kind. Rather than explain how the design of the George Washington Bridge obviates the kinds of suspension bridge failures that once plagued the form, he stresses the ultimate success of past projects and uses them to support the design that was going so far beyond experience.

Since the history of bridging the lower Hudson River between New York City and New Jersey had extended over many generations, there were naturally many unrealized proposals. Ammann recognized these politically and financially "unsuccessful attempts," but in this general paper he does not stress that there also were technical concerns and objections that were rooted in nineteenth-century and

early-twentieth-century bridge failures. Rather, he stresses successful designs and their aesthetic qualities and credits the success of
the George Washington Bridge design not only to his contemporary
colleagues but also to "those pioneers who by their vision and cour-
age have pointed the way, and whose ideas and studies have prepared
the ground for the final accomplishment." Although Ammann must
certainly have realized the important role that actual and imagined
failures had played in the pioneering efforts he so praised, he evidently preferred to stress the positive and only allude to the negative
in his report.

As Ammann relates, as early as 1868 efforts were made to span the
Hudson at New York, and by the early 1890s a cantilever structure with
a central span of about 2,000 feet had been proposed. The cantilever
would have been longer by 300 feet than the greatest span of the recently completed Firth of Forth Bridge, but the great success of that
structure gave considerable confidence to engineers that still longer
cantilevers could be built. Indeed, since the Secretary of War objected
to any bridge piers in the open water of the Hudson, a cantilever as long
as 3,000 feet would eventually be proposed, which would have essentially doubled the span as embodied in the existing state of the art in
one leap. This seems not to have bothered the various engineering
boards who reviewed proposed designs as much as the difficulties of
sinking a pier in the river and the enormous cost of a great cantilever.
It was only the failure of the Quebec Bridge while under construction
in 1907 that doomed once and for all such cantilever proposals.

A suspension bridge was from early on discussed as an alternative
to a cantilever, and the precedent of the Brooklyn Bridge and its
neighbors provided some inspiration for proponents. However, even
with these examples, the historical reputation of suspension bridges
required that they be considered with some caution. An early report
favoring the cantilever design exemplified the mood:

A suspension bridge spanning the North [Hudson] River without a pier
would involve such elements of uncertainty as regards first cost, novelty in

its magnitude as a hitherto untried engineering feat, and time of construction, to say nothing of the well-founded prejudice against the "suspension" principle for railroad purposes, as would render the enterprise impracticable from a financial standpoint.

Although Ammann (1933) quotes this passage, he does not comment directly upon it or its recitation of the reservations surrounding suspension bridges. Rather, he goes on to quote even more extensively the reports of an engineering board appointed by the Secretary of War and a subsequent engineering board appointed by the U.S. president, both of which were charged with looking into how long a span, of whatever design, could be safely and practicably erected over the river.

The Role of Economics and Aesthetics

Since the suspension principle was endorsed for a span of the order of 3,000 feet carrying as many as six railroad tracks, the Secretary of War ruled out a cantilever and thus paved the way for a suspension bridge. As early as 1887 the bridge engineer Gustav Lindenthal had proposed a suspension bridge to be located in lower Manhattan. The next forty years saw a host of different bridge designs proposed to cross the Hudson at different locations. In addition to the suspension and cantilever designs, an arch rising almost 600 feet above the water to span almost 3,000 feet across the river, a combination arch–suspension bridge, and numerous variations on purer suspension bridges with massively stiffened decks were proposed, as shown in Figure 9.1.

In his 1933 paper, Ammann discusses the many factors that went into choosing the type and location of a bridge, including traffic volume and patterns, topographical characteristics of the region, and geological factors. These latter considerations were so important for determining where piers could be located that, for

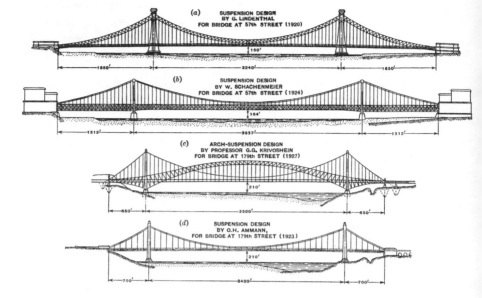

Figure 9.1. A selection of proposals to bridge the Hudson River (Ammann, 1933; by permission of the American Society of Civil Engineers).

Ammann, "it was a comparatively simple task to determine the type and general span arrangement of the main structure." However, even when it was finally determined in the 1920s that a suspension bridge should be erected at 179th Street in upper Manhattan, there was still room for considerable debate over the design. Where the piers were located determined the length of the main span of the bridge, of course, and Ammann argued for siting the piers 3,500 feet apart as the most economical choice. He dismissed the conventional wisdom that span length was the "predominant factor in the economy of a large bridge," pointing out the difficulties and uncertainties associated with putting a pier in deeper water to shorten the proposed span.

Ammann acknowledges in his report that the "unprecedented length of span . . . being exactly twice the longest suspension span in existence" raised the question among laypersons of feasibility.

However, he appeals to the authority of "engineers familiar with the design and construction of large bridges" and asserts unequivocally that "the span length and size of a bridge have nothing whatever to do with its safety, either during erection or after completion." He continues:

The feasible limit of span is reached when the metal required to carry a given load becomes excessive in cost and not because the safety is impaired. The physical limit of span is reached when no amount of metal can safely carry more than its own weight. The latter limit can be mathematically determined for the safe strength of any given material, and load conditions.

Such calculations, which Ammann reports were made under "conservative assumptions," led a board of engineers in 1894 to the conclusion that 4,335 feet was "the maximum span practical from an engineering point of view." In the intervening three decades, "accepted views regarding the proportioning of stiffening trusses in long spans" had led to much lighter-weight trusses than the Victorian engineers would have assumed, and high-strength steel wire was capable of carrying 50 percent more load while still maintaining a factor of safety of about two and one-half. Such evolutionary changes in analysis and materials enabled Ammann to conclude that "it may be demonstrated easily that a modern bridge of 10,000 ft. span could be built with perfect safety."

The maximum span of a suspension bridge having been declared to be three times what was proposed for the Hudson River crossing, the various hybrid designs could be dismissed on other grounds. In terms anticipatory of David Billington's (1983) "new art of structural engineering," Ammann makes his opinion of such impure designs clear:

Whatever may be the claims of scientific or economic merits of such hybrid types (and it is very doubtful in the writer's mind that they are justified because of the lack of structural simplicity and clearness of function of such incongruent systems), they cannot satisfy aesthetic principles.

Even a suspension bridge does not design itself, however, and the details of its form do not simply follow from its function. The human designer must make global decisions that set the stage for the local calculations that can then follow. Under the rubric of "conception of types of suspension system," Ammann presents the problem as follows:

Thus, while the type of bridge and its span arrangement logically forced themselves upon the designer, he was yet confronted with the more complex and controversial questions of selecting the appropriate form and proportions of the suspension system. . . .

In the case of the George Washington Bridge, the controlling criteria in selecting the system and its proportions were structural simplicity, maximum economy consistent with the required degree of rigidity, competitive conditions, and aesthetic conception.

The first criterion led to a system of a "plain cable with parallel chord stiffening trusses" as "unquestionably the simplest in its structural details." It remained to establish the nature of the stiffening truss. Ammann dismisses the heavier designs illustrated in Figure 9.1 as not "in conformity with the writer's conception of a graceful structure," and he elaborates as follows:

Furthermore, it is not necessary, in accordance with more recent views, to provide deep and rigid trusses in a long heavy span, and, particularly so, when the side spans are relatively short and the dead load relatively great, as in the case of the George Washington Bridge.

Extensive studies of the relative rigidity of similar structures and their behavior under actual conditions, and calculations of the degree of rigidity to be obtained in the selected system, led the writer to determine finally upon a very shallow and flexible truss, which not only resulted in far-reaching economy, but also effected a light and graceful appearance.

Ammann freely admits to being "influenced by his personal conceptions of taste" in the general design of his bridge, and he expresses a preference for bridges of the kind designed a century earlier to cross the Menai Strait. In what begins as an uncharacteristically personal passage in his generally objective-sounding report, Ammann reveals, albeit in the third person, how his own aesthetic model evolved:

> He has always been an admirer of the early English suspension bridges with their general simple appearance, their flat catenary, light, graceful, suspended structure, and their plain massive and, therefore, monumental towers.
>
> Deviations from the simple unstiffened cables were due to the efforts to give the system greater rigidity. This has been accomplished by various more or less efficient expedients, such as inclined stays. . . .
>
> It is significant, however, that after nearly a century of efforts to devise and introduce novel forms of suspension systems, or hybrids between the suspension type and other types, engineers, in designing the longest modern suspension bridges, have returned or adhered to the simple, naturally graceful forms which are characteristic of the early bridges of this type.

Ammann is remarkably impersonal in his allusions to historical bridges, mentioning neither Thomas Telford, who designed the Menai Strait Bridge, nor John Roebling, who adopted diagonal stays. More importantly, Ammann does not mention the troubles in the wind experienced by the early English bridges he so admired, nor does he mention that their light and graceful roadways were destroyed time and again by the wind. In contrast, as we have seen, when Roebling discussed these early bridges, he concentrated on their weaknesses and their inadequacies; Roebling devised diagonal stays, among other structural devices, to steady his own bridges against the destructive forces of the wind. By glossing over the root causes in failure of the evolutionary changes that had taken place in suspension bridge design throughout the nineteenth century, Ammann was inadvertently providing a model for twentieth-century designers that suppressed rather than highlighted the real problems that they faced in extending the envelope of experience.

Ammann goes on to discuss in more detail the question of stiffening trusses and wind bracing for the George Washington Bridge, and he admits that the "permissibility of an almost flexible system in the case of the completed bridge . . . was not obvious" when he began his studies. He criticizes the prior state of the art, in which a rigid stiffening system had come to be employed and the bridge had been analyzed with "the elastic theory, without respect to span length, dead weight of the bridge, or character of traffic." A more economical design resulted when the "correct or so-called deflection theory," which took into account the stiffening effect of the dead load of the flexible cables and the roadway itself, was employed. In support of the theory and its use in design, Ammann appeals to the proven application of a modified deflection theory by Leon Moisseiff, who had used it to design the Manhattan and Delaware River suspension bridges, and who was to design the Tacoma Narrows Bridge.

The wide deck of the George Washington Bridge, even as first erected as a single roadway that carried no rail traffic, did prove to be sufficiently stiff. The wind load was addressed by incorporating horizontal trussing, which was also designed to be relatively light because the heavy cables were expected to hold the bridge in place laterally. Ammann argued that the bridge would be perfectly safe even without a wind truss, but pointed out that one was incorporated into the design mainly to deal with unspecified "extraordinary wind effects." A more detailed report (Dana, Andersen, and Rapp, 1933) on the superstructure of the George Washington Bridge by three design engineers also paid little attention to wind loading, giving it a single paragraph in a sixty-page paper.

Although the dead weight of the bridge deck and cables was calculated to keep them from deflecting sideways in the wind, the pressure of the wind along the 3,500-foot span did have to be reacted against by the towers. The structural design of the towers was one thing, however; their aesthetic design was another, and Ammann's 1933 paper dwells more on the latter. He again admits it to be largely a matter of taste as to what exact form the towers would take:

It is futile to theorize about this question – it is largely a matter of aesthetic conception, which is so intensely individual and changeable – nor can it be dealt with on general principles without regard to the local economy or landscape. Moreover, the aesthetic treatment of a bridge, as that of any other engineering structure, is not always satisfactorily solved even by correct and honest application of engineering principles.

In discussing whether the steel towers should be encased in "an architectural treatment" of stone, Ammann apparently forgets the "clearness of function" principle by which he dismissed hybrid bridge designs and says that he is "not impressed with the criticism, based solely on theoretical and utilitarian grounds, that the encasement would constitute a camouflage which would hide the true structure and its function." Although economics may later have influenced the decision to leave the steel framework towers bare (see Fig. 9.2), clearly the superficial appearance of his bridge was at least as important to Ammann as its structural integrity, and a repeated appeal to aesthetics and personal design preference strongly flavors the engineer's report. There seems little doubt that Ammann's articulate defense of the designer's aesthetic prerogative was embraced by his contemporaries, as the light and graceful appearance of suspension bridges built throughout the 1930s in the United States attests.

It was thus in this design climate, where questions of economy and silhouette competed with those of factor of safety and rigidity, that ever longer, slenderer, and lighter suspension bridges were being conceived and erected. This is not to say that structural engineering analysis was neglected, for the concerns for more "correct" theories show that clearly not to have been the case, and Ammann's George Washington Bridge stands today as a model of bridge engineering. However, the appeal to history primarily for aesthetic models, coupled with the hubris of analytical sophistication that the pioneers clearly did not possess, appear to have led bridge designers to forget or dismiss as irrelevant the behavior and failure of early

Figure 9.2. The George Washington Bridge with a single deck, as originally opened in 1931 (The Port Authority of New York and New Jersey).

suspension bridges in the wind. The static analysis of wind effects, rationalized by observations about the "inertia of the enormous dead weight resisting sudden gusts of wind," took place in a total design climate that saw early bridges as light and pretty structures to be emulated. Their flaws and failures appear to have been effectively forgotten.

The Tacoma Narrows Bridge

Burt Farquharson, a professor of civil engineering at the University of Washington, was studying a model of the Tacoma Narrows Bridge on the Seattle campus and was observing the behavior of the actual bridge across the Narrows in the months between its completion and collapse in 1940 (see Figs. 9.3 and 9.4). He opened his series of reports (1949) with a brief but broad historical review of the dynamic behavior of suspension bridges. In this survey he notes that the failure of the contemporary bridge "came as such a shock to the engineering profession that it is surprising to most to learn that failure under the action of the wind was not without precedent." Such an observation reinforces the interpretation of the design climate, as exemplified by Ammann (1933), to be one in which successful precedents were emphasized over failures and in which successive advances in analysis emphasized the importance of analysis per se over the focus on failure modes that earlier bridges demonstrated.

As Farquharson reports, in the seven decades between 1818 and 1889, ten suspension bridges were severely damaged or destroyed by the wind. Table 9.1, taken from Farquharson (1949), summarizes the historical record and includes the Tacoma Narrows for comparison. Many of the failures that occurred before 1840 were discussed by Roebling (1841) in his paper describing what he as a suspension bridge designer needed to know in order to design a successful bridge (see Chapter 8). In his Niagara Gorge Suspension Bridge, finished in 1855, Roebling not only built on the experience of others but also built the first suspension bridge to carry railroad traffic, thus providing a counterexample to the conventional design wisdom that the suspension bridge was not suitable for such heavy and concentrated loads. However, Roebling's model of good practice, which rested upon failure awareness, analysis, and avoidance, and the design logic inherent in a failure-based approach, were de-emphasized, if not completely forgotten or unappreciated, by succeeding generations of bridge designers.

According to Farquharson, in the five decades between the failure of the Niagara-Clifton Bridge in 1889 and the Tacoma Narrows, "the building of suspended spans of ever increasing length and load-carrying ability marched steadily forward." Furthermore, since vehicular rather than rail traffic became the design objective, there was

Figure 9.3. The Tacoma Narrows Bridge in its fatal torsional oscillation mode, 1940 (University of Washington).

a "removal of the exacting requirements of a controlled grade-change," which resulted in "much lighter stiffening structures." However, in concentrating on the stiffness under the traffic load, designers were overlooking the role of stiffness in suppressing oscillations of the bridges in the wind, a failure-based design objective that Roebling considered equally with stiffness under moving traffic loads.

By the time the George Washington and Tacoma Narrows bridges were being designed, however, the effect of the wind was considered seriously only in its effect on deflecting the roadway laterally (cf. Ammann, 1933; Dana, Andersen, and Rapp, 1933). According to Farquharson, even years after the collapse of the Tacoma Narrows Bridge it was "common practice . . . to consider a structure adequately prepared to withstand the onslaught of any wind provided only that it was designed for a static wind pressure of 30 lb per sq ft of projected area or, in the case of a bridge, 1½ times the area as

Figure 9.4. The Tacoma Narrows Bridge collapsing, 1940 (University of Washington).

seen in elevation." In a footnote he calls this requirement "the lineal descendant of the 56 lb which grew out of the failure of the Tay bridge [which was blown off its piers in the wind] in 1880 [*sic;* actually 1879]," a reasonable observation given the propensity of the design community to lower performance requirements and factors of safety when analysis advances in the absence of failures, as characterized the sixty years of bridge building between the Tay disaster and the Tacoma Narrows collapse. The absence of failures in a particular mode (in this case, wind-induced) tends to lead designers to forget the first principles of designing against that mode of failure. However, as Farquharson so correctly points out:

The successful performance of these structures with relation to the wind should not be too lightly attributed to their weight or stiffness character-

Table 9.1. *Suspension Bridges Severely Damaged or Destroyed by Wind*

Bridge (location)	Designer	Span (ft.)	Failure date
Dryburgh Abbey (Scotland)	John and William Smith	260	1818
Union (England)	Samuel Brown	449	1821
Nassau (Germany)	Lossen and Wolf	245	1834
Brighton Chain Pier (England)	Samuel Brown	255	1836
Montrose (Scotland)	Samuel Brown	432	1838
Menai Strait (Wales)	Thomas Telford	580	1839
Roche-Bernard (France)	Le Blanc	641	1852
Wheeling (United States)	Charles Ellet, Jr.	1,010	1854
Niagara-Lewiston (U.S.A.–Canada)	Edward Serrell	1,041	1864
Niagara-Clifton (U.S.A.–Canada)	Samuel Keefer	1,260	1889
Tacoma Narrows (United States)	Leon Moisseiff	2,800	1940

Source: Farquharson (1949).

istics considering the paucity of detailed information regarding the winds to which they were subjected. . . . That significant motions have *not* been recorded on most of these bridges is conceivably due to the fact that they have never been subjected to optimum winds for a sufficient period of time.

It is certainly not recommended design practice for engineers to concoct inconceivable loadings, but Farquharson's point embodies an essential rule of design *logic:* the absence of failure does not prove that a design is flawless, for a latent failure mode may be triggered by yet-unexperienced conditions. However, it appears to be a trait of human nature to take repeated success as confirmation that everything is being done correctly. In such a design climate each successive success reinforces this view and drives the state of the art further from considering fundamental failure criteria as the true principles upon which designs must be based.

The absence of dramatic failures can not only make designers complacent with regard to the genre of which they are so justifiably proud; a climate of success can also make designers react more slowly to warning signs that something is wrong. Even Farquharson, who was personally observing so closely not only the actual Tacoma Narrows Bridge but a model of it for three months, evidently did not expect the bridge to fail catastrophically the way it did. As he points out, even before the Tacoma Narrows Bridge was completed, there were other bridges built to the same state of the art that were oscillating in the wind, as summarized in Table 9.2. Although no motions observed in these bridges prior to 1940 were considered "very significant," in the wake of the collapse of the Tacoma Narrows Bridge their motions came under greatly increased scrutiny. Rather than circulating as mere rumors, "certain conditions surrounding some of the most recent bridges were more fully reported and discussed." The Bronx-Whitestone Bridge, which had exhibited some flexibility after its completion in 1939 (see Fig. 9.5), was eventually stiffened with a distracting truss (see Fig. 9.6) that interferes with what was once an excellent view of the Manhattan skyline.

Moreover, the collapse of the Tacoma Narrows Bridge brought the long historical record out of the pages of books on the history of engineering and technology, generally believed to be written and read largely by engineers in retirement of one form or another, into the pages of *Engineering News-Record* and other publications of mainstream engineering (see, e.g., Finch, 1941). The parallels between the failure of the Tacoma Narrows Bridge and its century-old counterparts like the Menai Strait and Brighton Chain Pier suspension bridges were undeniable (cf. Fig. 8.4). Certainly the state of the art had advanced considerably since that time, with bridge spans having

Table 9.2. *Modern Suspension Bridges that Have Oscillated in the Wind*

Bridge (location)	Year built	Span (ft.)	Type of stiffening
Fykesesund (Norway)	1937	750	Rolled I-beam
Golden Gate (United States)	1937	4,200	Truss
Thousand Islands (U.S.A.–Canada)	1938	800	Plate girder
Deer Isle (United States)	1939	1,080	Plate girder
Bronx-Whitestone (United States)	1939	2,300	Plate girder
Tacoma Narrows (United States)	1940	2,800	Plate girder

Source: Farquharson (1949).

been extended by almost an order of magnitude (see Fig. 3.3), but the modes of failure against which the oldest bridges had to be designed are modes of failure against which even the newest bridges today must be designed. When the design climate becomes such that this elementary principle is lost sight of, as it appears to have been in suspension bridge building in the early to mid-twentieth century, then successful genres appear almost doomed to evolve (or devolve) toward a colossal failure.

Conclusion

Although the circumstances and details of the design climate that prevailed when the Tacoma Narrows Bridge was planned and built may be unique to suspension bridges, the story of that bridge reveals the myopia that can occur in the wake of prolonged and remarkable success and that is endemic to the design process itself. The stories

Figure 9.5. The Bronx-Whitestone Bridge, as built in 1939 (Triborough Bridge and Tunnel Authority).

of the Dee, Tay, and Quebec bridges, for example, follow very similar scripts, *mutatis mutandis,* as can be seen so clearly in the extended studies of Sibly (1977; cf. Sibly and Walker, 1977; Petroski, 1993), which are discussed in Chapter 10.

Concerns with aesthetics and what might generally be considered softer criteria for design do not have to prevail at the expense of hard criteria like strength and stability. There are countless examples of extremely successful, novel, and daring structural designs that are aesthetic successes as well. There are differences of opinion regarding the structural honesty, economy, and other characteristics of bridges like the Britannia, Brooklyn, and Firth of Forth, but these are incontrovertible successes in the realm of structural design, and they can all be admired, at least by their champions, on aesthetic grounds also.

Figure 9.6. The Bronx-Whitestone Bridge, as modified in 1946, employing the original stiffening plate girders as the bottom chord of an unattractive stiffening truss (courtesy of the Triborough Bridge and Tunnel Authority).

Successful structures do not follow simply and directly from an engineer's conceptual sketch or an architect's model, no matter how appealing visually. Asking the right questions regarding stiffness, strength, stability, and more is the surest way of achieving structural success, regardless of whether the artifact is considered beautiful or ugly. Aesthetic considerations alone obviously will not ensure a safe bridge, but neither will a careful structural engineering analysis dictate an unattractive one. Structural and aesthetic considerations are simply different, and not necessarily always separate or incompatible, aspects of the same design process. However, without a hard, correct, and complete structural analysis, even the most beautiful of bridges can be at risk.

No matter when designed and built, a bridge (or any other structure or machine) must obey the laws of nature and withstand the forces of nature. Since natural laws are fundamentally immutable and natural environments and climates change gradually if at all, every bridge built today is potentially subject to the same failure modes as every bridge built in the past. Those bridges that have been structural successes do not readily reveal in their own aesthetic lines or even in their distinct technical specifications the secrets of their success, hence, extrapolation from them alone is done at the designer's risk.

The more meaningful design information contained in case histories of successful projects is: what failure modes were anticipated, what failure criteria were employed, and what failure avoidance strategies were incorporated. Nineteenth–century designers like John Roebling were often quite explicit about such aspects of failure, and structurally daring suspension bridges like the Niagara Gorge and Brooklyn are testaments to the value of proactive failure analysis in design. The apparent ignorance, displacement, or casualness of such a failure–based principle, as exemplified by the apparent preoccupation of designers a half–century after Roebling with such criteria as aesthetics and economy, created a design climate that culminated in the collapse of the Tacoma Narrows Bridge.

When the prevailing design climate recognizes case histories only to justify extrapolations to larger and lighter structures, a form of normal engineering may be said to be practiced. However, when such normal and ordinary engineering is being used to design and build structures of an abnormal and extraordinarily large scale, the lessons of the abnormal and extraordinary – as embodied in case histories of failures and how to use them – are much more relevant. The greatest engineers of the past seem intuitively to have recognized this in designing and building their truly revolutionary structures, and it behooves modern designers to look back at the mistakes of the past and how they were dealt with in addition to looking for precedents of success.

10

Historic Bridge Failures and Caveats for Future Designs

The relationship between success and failure in design constitutes one of the fundamental paradoxes of engineering. The accumulation of successful experience tends to embolden designers to attempt ever more daring and ambitious projects, which seem almost invariably to culminate in a colossal failure that takes everyone by surprise. In the wake of failure, on the other hand, there is generally a renewed conservatism that leads to new and untried design concepts that prove ironically to be eminently successful precisely because the design process proceeds cautiously from fundamentals and takes little for granted. As the new design form evolves and matures, however, the cautions attendant upon its introduction tend to be forgotten, and a new period of optimism and hubris ensues. This cyclic nature of the engineering design climate has been elaborated upon here and elsewhere and is supported by numerous case studies (see Petroski, 1985).

In his thesis on structural accidents and their causes, Sibly (1977) analyzed several large metal bridge failures in a design-historical context and showed that each of them occurred in a design climate characterized by increasing span length, increasing slenderness, increasing confidence in analysis, or decreasing factor of safety (cf.

Sibly and Walker, 1977). Generally a combination of such circumstances exists and there are typically warning signs of impending trouble, but they are seldom if ever paid very explicit attention in the evolutionary design process. Following a failure, however, such trends become very evident in retrospect. In the discussion at the end of their reflective paper, Sibly and Walker conclude that common features were present in the design circumstances leading up to the several major bridge failures that their study comprised:

In each case one can identify a situation where, in early examples of the structural form, a certain factor was of secondary importance with regard to stability or strength. With increasing scale, however, this factor became of primary importance and led to failure. The accidents happened not because the engineer neglected to provide sufficient strength as prescribed by the accepted design approach, but because of the unwitting introduction of a new type of behavior. As time passed during the period of development, the bases of the design methods were forgotten and so were their limits of validity. Following a period of successful construction a designer, perhaps a little complacent, simply extended the design method once too often.

Understanding how error and oversight have led to failures in the past provides a model for the critical examination of present practice. Such an examination in itself may provide a self-correction to the design process so that the "inevitable" failure may be prevented, or at least postponed – perhaps indefinitely. Given the evidence assembled by Sibly and Walker and the apparent cyclic nature of success and failure in design, it is natural to ask if there are any indications of impending failures in designs on the drawing board today. What are the most likely engineering projects to be accident-prone? Is it possible to look at current design practice in context and see patterns developing that point to trouble ahead? Such questions can and should be asked, of course, for their very articulation can lead to a process of reflection that might correct an otherwise ill-fated trend in design.

Extrapolating from Sibly and Walker's Pattern

Case histories studied by Sibly and Walker (1977) are listed in Table 10.1. As they note, "more as a point for discussion than as a serious observation," the Dee, Tay, Quebec, and Tacoma Narrows bridge failures occurred very nearly thirty years apart. Although Sibly in his thesis (1977) and he and Walker in their paper (1977) did not study in depth the series of accidents that occurred in steel box-girder designs, exemplified by the 1970 incident with the Milford Haven Bridge in Wales, Sibly and Walker observe that this provides further confirmation of the existence of a thirty-year cycle, which they freely admit could merely be coincidence. Whether coincidence or not, the striking pattern noted by Sibly and Walker seems inevitably to call attention to the years around the turn of the millennium as a period during which a major bridge failure of a new kind might be expected to occur.

Sibly and Walker speculate that the thirty-year interval may represent a "communication gap between one generation of engineer and the next," and this may indeed be involved in the explanation. New bridge types do not suddenly appear fully constructed, of course, and it can take decades for a concept to go from the drawing board to the field. However, a new bridge type will generally not become widely specified until a pioneer design is well under construction, if not completed. Engineers attempting a new or relatively new design will take special care to reason from first principles and try to anticipate all modes of failure. It will only be after a bridge type has become "standard" that first principles will tend to be forgotten, that less and less thought will be given to extrapolating incrementally from past experience, and that more and more design decisions will be delegated to less and less experienced engineers. In such a situation, it is likely that generational gaps may indeed develop.

Individual bridge failures do occur more frequently and irregu-

Table 10.1. *Landmark Bridge Failures*

Bridge (location)	Type	Probable cause	Year	Interval (years)
Dee (England)	Trussed girder	Torsional instability	1847	—
Tay (Scotland)	Truss	Unstable in wind	1879	32
Quebec (Canada)	Cantilever	Compressive buckling	1907	28
Tacoma Narrows (U.S.A.)	Suspension	Aerodynamic instability	1940	33
Milford Haven (Wales)	Box girder	Plate buckling	1970	30
?	Cable-stayed?	An instability?	c.2000	c.30

Source: Sibly and Walker (1977).

larly than every thirty years, of course, but the pattern identified by Sibly and Walker speaks more to the failure of a major bridge type than to a particular bridge. Indeed, each of the successive bridge types listed in Table 10.1 may be argued to have developed to the point of failure precisely because it offered an alternative to the previously failed types. Trussed girders became obsolete after the Dee Bridge failure, and iron trusses evolved in their place. Following the Tay accident, the cantilever design became popular for railroad bridges until the Quebec collapse. In the wake of that failure, suspension bridges became favored over cantilever designs until the Tacoma Narrows collapse. In looking to what bridge type might be the focus of the next major failure, we should direct our attention to currently competing and evolving bridge types.

What Drives Design

There are many obvious factors that cause larger and larger structures to be designed and built, and the phenomenon is as old as engineering itself. As we have seen in Chapter 3, Vitruvius related the story of how the engineer Diognetus lost his yearly contract to defend the city of Rhodes when another engineer, Callias, promised in a lecture to make a bigger and better crane for capturing enemy siege machines. Bridge types have continued to evolve to longer and longer lengths for obvious reasons, not the least of which is the fact that there seem always to be engineers who want to build longer bridges. Indeed, in the last century, Baker (1873) pointed out in his book on long-span bridges that "the size of a bridge is too commonly the popular standard by which the eminence of its engineer is measured."

In a recent article (Fowler, 1988) on the lengths to which suspension bridges can be taken with new materials like carbon-fiber composites for their cables, it is pointed out that "the history of long spans is the history of materials" and that "each new material took 30 to 40 years to gain engineers' acceptance and confidence." The British firm of G. Maunsell & Partners wished to exploit the newest material first and was "looking for a client and somewhere to build" a footbridge as the first suspension structure to employ composite materials. According to one of the firm's engineers, "We'd treat it as a research project and monitor the performance of all aspects over a 10 to 20 year period. The technology is available to build such a structure. All the nuts and bolts are there – we'd like to do it as a world first." Should such a project succeed – and, with such close attention paid to its inception and such close monitoring of its behavior and aging, it could certainly be ex-pected to succeed – the proponents of the footbridge design would hope to develop the five-mile (eight-kilometer)* spans that theoretical cal-culations suggest are possible.

*In keeping with the style of using the historical British system of units in this book, more recent and planned suspension and cable-stayed spans will be given primarily in feet, with parenthetical conversions to meters.

As new structural types and materials become incorporated into practice, a desire for more ambitious and also more economical designs seems invariably to come. The objective of economy is commonly achieved by reducing factors of safety. A recent article (Kuesel, 1990) asking "Whatever happened to long-term bridge design?" cautioned that "we should not be debating how much further we can reduce safety margins (for loads and strength), but rather how much we should increase safety margins (for wear and corrosion)." This is especially sage advice for structures that are expected to last a century or more.

Cable-stayed Bridges

The cable-stayed bridge is without question the most significant new type currently being developed, and it is evolving as a strong competitor for long-span applications. Though sometimes labeled "exotic," cable-stayed designs have the distinct advantages of allowing construction without significant falsework while at the same time maintaining a slender roadway profile. There were still relatively few cable-stayed bridges in the United States by 1990, and engineers and contractors retained a reluctance to face their challenges. The new designs were still "lumped into the category of exotic bridges that scare the hell out of contractors," for "you can't build them from a set of plans" (Anonymous, 1992a).

Yet, for all the newness of the structural type, a contemporaneous article (Takena et al., 1992) acknowledged that "more than 100 years of service life are demanded" of cable-stayed bridges and pointed out that the competing objectives of fewer and more highly stressed cables and longer life were clearly at odds, for "as the diameter of a cable increases, its fatigue strength decreases." Elsewhere (Bruneau, 1992), it was observed that "it is already recognized and anticipated that the engineering efforts required for the design of cable-stayed bridges exceed those required for more standard bridge configurations." Among the complications of analysis are the structural combinations that must be considered to allow for a cable being missing, perhaps due to a fatigue failure, or being in need of replacement. At least one bridge designer described "the prospect of analysing all the possible combinations of structure" as "frightening" (see O'Neill, 1991).

Although cable-stayed bridge designs date at least from the Renaissance (Fig. 10.1 shows a proposed design from the late sixteenth century), the modern history of the form is usually reckoned from the work of the French civil engineer Claude-Louis Navier in the early nineteenth century. The present popularity of the genre is a result of post–World War II rebuilding in Germany (see Fig. 10.2), as described recently by Billington and Nazmy (1990). Over forty major cable-stayed bridges were built around the world before the first one was built in the United States, in 1973 in Alaska. (The first cable-stayed bridge built in the contiguous United States was the Pasco-Kennewick Bridge erected across the Columbia River in Washington in 1978.) Britain did not get a major modern cable-stayed bridge until the Dartford Bridge across the Thames was finished in September 1991 (at which time it was christened the Queen Elizabeth II Bridge); only five and one-half years elapsed "between the germ of an idea and completion" of the new bridge (Byrd, 1992).

According to Billington and Nazmy, in the United States as in Germany, in part, "cable-stayed designs initially got built because designers wanted to use the new form and were attracted to it aesthetically." But the "aesthetic problem" of achieving "a light deck by means of many closely spaced cables that do not appear visually cluttered" is, like many aesthetic problems, more easily stated than solved. Cable-stayed bridges have had their aesthetic advocates and detractors, of course, but even the most ardent of supporters would most likely agree that not all designs have been equally successful. Readers can judge for themselves in an appendix to Billington and Nazmy's paper, which "summarizes the major cable-stayed bridges completed between 1955 and 1987" by showing graphically eighty-

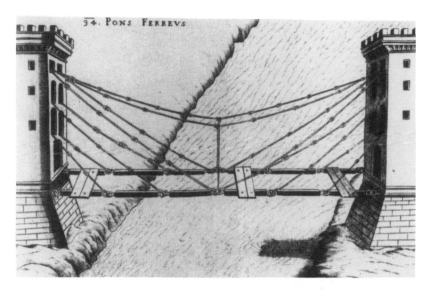

Figure 10.1. A cable-stayed bridge design from Fauston Veranzio's *Machinae Novae* (1595).

five variations that designers have come up with for supporting a light deck by cables.

The design trend of increasingly large cable-stayed bridges is typical of the historical trends in the bridge types studied by Sibly and Walker (1977). A Japanese cable-stayed bridge with a 1,608-foot (490-meter) span was completed in the early 1990s, as was a record-setting Norwegian span of 1,739 feet (530 meters). This record was expected to stand for only about two years, until the French would bridge the mouth of the Seine with a cable-stayed design of 2,808-foot (856-meter) span. Cable-stayed spans in excess of 4,000 feet (1,200 meters) have been put on the drawing board, and longer ones can certainly be conceived. Figure 10.3 shows the trend of maximum span length of the cable-stayed bridges tabulated in a survey paper by Podolny (1992). Conventional suspension bridges generally continue to present the more economical alternative for spans in the range of 5,000 feet (1,500 meters), but there is no clear early indi-

Figure 10.2. A group of cable-stayed bridges designed for Düsseldorf (Leonhardt, 1984; by permission of the author).

cation that the growth curve of cable-stayed designs will flatten out at that length.

As long as the cable-stayed bridge remains an "exotic" form, it can be expected that engineers and designers will treat it and its possible failure modes with caution and use large factors of safety. Indeed, increasing maximum span length alone is no certain harbinger of disaster, as the Firth of Forth Bridge so clearly demonstrated. When that bridge's 1,710-foot (521-meter) cantilevered spans were completed in 1890, they represented a full doubling of the maximum span then known. The first Quebec Bridge, which collapsed during construction in 1907, was to have had a much lighter cantilevered span of approximately the same magnitude, which by then represented a leap of only 50 percent beyond the contemporary limits of practice (see Sibly and Walker, 1977). The eminently successful

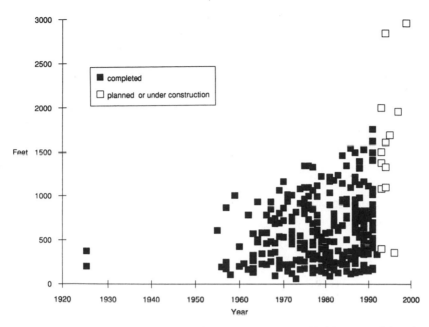

Figure 10.3. The completion dates and length of cable-stayed bridge spans (plotted by Tonya Dale, from data in Podolny, 1992).

Brooklyn and George Washington suspension bridges each doubled the maximum existing span of their times (see Fig. 3.3).

What the historical record of bridge failures cautions us against is not so much sheer size as large bridges designed and built in a climate of overconfidence. At the present time, cable-stayed bridge designs appear still to be held in sufficient respect so that there is little reason to be concerned for their immediate safety. However, as the form becomes more commonplace and as those who pioneered its design and construction retire from practice, there is reason to fear that the design climate in which long-span cable-stayed bridges are conceived and analyzed will be at least one professional generation removed from fundamental design concerns. It is when this happens that a catastrophe might be expected. A harbinger of this occurred in mid-1992, when a 394-foot (120-meter) cable-stayed concrete bridge over the Han River in Korea collapsed during construction. However, since the accident occurred before any cables were stressed and was attributed to "poor quality [concrete construction] work" (Anonymous, 1992b), designers may not pay special attention to the failure. Yet such seemingly dismissible precursor incidents occurred in case after case of major structural failures studied by Sibly (cf. Sibly and Walker, 1977).

In summary, at its present rate of development, the genre of cable-stayed bridges could present a prime candidate for a colossal failure around the year 2000 for several reasons: (1) the long-span cable-stayed bridge can be expected to be commonplace; (2) pioneers of this bridge type are no longer as influential in designing or consulting on the form; and (3) with scores of successful bridges standing for decades and supporting their case, commissioners and designers of new cable-stayed spans can be expected to move toward lighter, slenderer, more economical, and more daring spans. However, there is no reason to think, especially with an awareness of the caveats, that failure is inevitable (cf. Petroski, 1993).

The Argument Against Predestined Failure

The cable-stayed form appears to be a prime candidate for continuing the thirty-year pattern suggested by Sibly and Walker's (1977) analysis, but there are also circumstances that mitigate against a major cable-stayed failure around the year 2000. Ironically, it is the strong pattern pointed out by Sibly and Walker that, if widely acknowledged and discussed within the design community, may itself present circumstances that break the pattern. As engineering designers become more aware of historic cycles of success and failure, and as they see the pitfalls of overconfidence and complacency, their designs can be expected to be carried out with renewed care and attention to detail. The perspectives on failure offered by work such as Sibly and Walker's, as summarized in Table 10.1, make it reasonable to expect that one should be able to anticipate potentially dangerous design climates and should be better able to obviate failure. The preponderance of landmark failures caused by instabilities makes it increasingly apparent where designs should be scrutinized for at least one kind of potential surprise. Sibly, in his thesis (1977), further argues for ongoing design reviews of evolving structural types as one very cost-effective means of identifying developing problems.

The superficial structural similarity between suspension and cable-stayed bridges places the latter form in a historically special position. As long as the dramatic filmed record of the Tacoma Narrows Bridge collapse remains very high in the consciousness of bridge engineers, it serves as a constant reminder of the potential frailty of large suspended structures of all kinds. However, there is some reason to fear that if discussions of the Tacoma Narrows collapse are left to physics textbooks and applied mathematicians, as recently pointed out by Billah and Scanlan (1991), then the engineering significance of the failure may be more removed from practice than engineers should like. The failures of the past hold lessons for engineering that are much more profound than simply as examples of physical phenomena or solutions to differential equations. It is important, therefore, that engineering curricula do not leave analysis of such accidents as the Tacoma Narrows collapse entirely to the early engineering courses. Rather, it behooves the profession to elaborate on case studies of failures throughout the curriculum, and especially in the later and more advanced design courses.

A strikingly common feature of the failures studied by Sibly and Walker, and summarized in Table 10.1, is the presence of one form or another of instability as a probable cause. This is consistent with the observation that the accidents followed a period of consciously increasing scale and slenderness ratios that in turn unwittingly brought into primary importance factors that once were of secondary importance at best (cf. Chapter 3). In the cases studied by Sibly and Walker, the ill-fated designs were generally carried out in a climate of confidence, in that each bridge that failed either represented an evolutionary increment in a long line of similar bridges that had performed satisfactorily or represented a step taken with the assurance that it was being done with an adequate factor of safety or with a sound foundation in theory. It may take special efforts for designers engaged in such pursuits to look at their designs with the perspective of the work of Sibly and Walker, but the case for doing so is compelling.

Conclusion

Sibly and Walker's observation that over the past century and a half major bridge failures have occurred every thirty years or so may in the long run prove to be little more than an oddity or curiosity, but it also may serve to alert engineers to some warning signs in the nature of the design process that can have profound implications for the evolution of large structures of all kinds. Although the bridges in Table 10.1 are of vastly different structural types and their failure modes are ostensibly distinct, there is an underlying sameness to the design process that brought each of the bridge types to the point of collapse. Indeed, there are fundamental features of the engineering design process that are so timeless and inextricably a part of the creative aspects of design that even the most sophisticated developments of artificial intelligence and computer-aided design are unlikely to alter the basic fact that only human insight can eliminate human error from design.

In John Hersey's (1957) novel about a young engineer traveling on a Chinese junk up the Yangtze River, the engineer expresses surprise that there could be a new way to negotiate dangerous rapids in a boat whose form over the ages had changed as little as the river itself. The old boat owner assures him that such is the case, however: "There is a new way every forty years. It is an old way but it is called new. River pilots wait only for their grandfathers, who know better, to die before they claim they have found a new way." Western engineers may only wait thirty years, but they too are frequently calling old ways new (and still new ones old) and forgetting what their professional grandparents and older ancestors in design understood. Well over a century before the Tacoma Narrows collapse, there were suspension bridges and wind-induced failures of them. Over a century before steel box-girder bridges failed, Robert Stephenson built wrought-iron tubular bridges that were successful precisely because he stiffened their plates to avoid buckling. And cable-stayed bridges were conceived by Navier well over a century before the present enchantment of engineers and bridge commissioners with the new form.

Sibly and Walker's thirty-year pattern, which transcends even the behavioral problems characteristic of individual bridge types, may be dismissed as a remarkable coincidence, but there is every reason to believe that ignoring the implications of it and the lessons that can be drawn from such a perspective on structural failures may lead ironically to a new entry in the table. The surest way to obviate new failures is to understand fully the facts and circumstances surrounding failures that have already occurred. Sibly and Walker have observed an order in the mistakes of the past, and the greatest lasting value of work such as theirs may lie in interrupting that kind of order.

11

Conclusion

The development of failure-based paradigms can serve to define better the role of failure in the design process. Understanding how error has been introduced into past designs makes it easier to identify and eliminate it in future designs, and demonstrating how error has been consciously avoided by some of the most creative designers of the past provides models of good design for the future. When we understand both the negative and positive aspects of the role of failure in the design process, the process itself can be made to be more understandable, reliable, and productive.

Paradigms such as those presented here can also serve to unify a wealth of past experience of failure and failure avoidance that has generally escaped useful classification schemes. The value of case histories has long been appreciated, but the systematic use of them to benefit the next generation of designs can be elusive. By providing a new structure of paradigms within which case studies can be given broad interpretations across design specialties, a wealth of new experience is made available to design theorists and practitioners alike.

Design has been a notoriously problematic aspect of the engineering curriculum, and paradigmatic case histories provide a means of understanding and teaching the design process that can be structured according to a beneficial scheme. The scheme of paradigms and supporting case studies has the dual advantage not only of laying the foundation for organizing the mass of design experience for theoretical studies but also of providing caveats and models for the student and practitioner of design.

Since human error is incontrovertibly a major cause of design failures, understanding how error-prone situations can be identified in the design process helps to reduce error. With the reduction of error we can expect broad social and technological benefits ranging from increased reliability of engineered artifacts and systems to increased productivity and competitiveness in the world marketplace.

If valid paradigms have been identified and presented here in such a way that the reader can supplement the supporting case histories for a particular paradigm with additional ones from his or her own field of experience, then the hypothesis of that paradigm's validity will be confirmed. If the student engineer or practicing designer can read this book and can use it as a model for approaching other engineering problems or catching design errors in his or her own work in progress, then the hypothesis of useful design paradigms will have been validated by their productive use in increasing the effectiveness of engineering training and practice and in the increased reliability of engineering design.

Case studies clearly complement any abstract principles of design, for in concrete examples are embedded models of error to be avoided and of good practice to be emulated. The explication and dissemination of a collection of case studies can thus provide the basis of a common core of design examples from which the nature and classification of kinds of human error and steps that have been taken against them may be understood. Case studies drawn especially from the classical engineering literature provide insights into the nature of human error and its role in the design process that, because their interpretation is generally not clouded by issues of current technical, legal, economic, or social debate, are more likely to be objectively presented, interpreted, and received by the design community.

The lessons of cogent case histories are timeless, and hence a heightened awareness among today's designers of classic pitfalls of design logic and errors of design judgment can help prevent the same mistakes from being repeated in new designs. Although our most modern scientific, mathematical, and computational tools are without precedent in the history of design and manufacturing, there are essential elements of the design process that are more fundamental than any of the modern tools of or perspectives on design. Thus, lessons learned from errors made in those fundamental aspects of the design process are no less relevant today than they were when they were made. By inculcating in tomorrow's engineers a sense of error and error avoidance in the context of case studies, we can expect the next generations of designers at least to be forewarned about what mistakes have been made and how they might be avoided in the future.

The collection of case histories in this book is in no way meant to be definitive. Rather, this collection is intended to demonstrate the potential of employing case studies to illuminate how human error has been committed and avoided by engineers throughout history. The long historical perspective emphasizes the persistence of error in the design process, and the clear association of mistakes with the cognitive aspects of design provides a strong warning about the growing use of computer-based tools in design. The development of computer aids for design and other aspects of engineering cannot be expected to be free of error, and denying, de-emphasizing, or ignoring this fact can only create a climate even more hospitable to error than one in which it is high in the consciousness of those working on engineering design problems or on the development of computer software to attack them. Computer-aided design tools created in a methodological vacuum devoid of past experience are likely to provide more than their own fair share of case studies of error and failure for the next generation.

Although, as discussed in the Introduction, this book is dominated by structural engineering examples, they do demonstrate a range of overlapping types of error that have dogged the design process from earliest times. Rather than diffuse the effect of this collection by reaching for examples from the various branches of engineering, the design principles embedded in the structural examples have been presented as paradigms for design generally. Readers who are not inclined toward structural engineering should be able to call up analogous examples from their own more familiar, albeit more recently developed fields.

The Bibliography in the back of this book is also heavily oriented toward the structural engineering literature, but this is not to say that they should be seen as limited in their relevance or restrictive in their lessons. The purpose of this book is to present an incontrovertible paradigmatic argument for the value of case histories, ranging from the ancient to the modern, in illuminating and eliminating the causes and results of human error in the design process generally.

The Relevance of History to Engineering

The human activity of engineering design is not a perfect science capable of producing perfect products. Engineering is part art, and it is this aspect of design that is difficult if not impossible to quantify and model completely. There is no finite checklist of rules or questions that an engineer can apply and answer in order to declare that a design is perfect and absolutely safe, for such finality is incompatible with the whole process, practice, and achievement of engineering. Not only must engineers preface any state-of-the-art analysis with what has variously been called engineering thinking and engineering judgment, they must also supplement the results of their analysis with thoughtful and considered interpretations of the results.

Success can be replicated virtually without risk, but it can be extrapolated only by a proper application of the engineering method embodied in a proper perspective on failure. Engineering advances by proactive and reactive failure analysis, and at the heart of the engineering method is an understanding of failure in all its real and imagined manifestations. Improvements in analytical tools and models cannot alone improve the practice of engineering and the reliability of its products, for the choice of the input to and the interpretation of the output from analysis involve extrascientific judgment. Indeed, efforts to improve engineering design by concentrating on the refinement of its more easily quantifiable analytical models and tools may actually be counterproductive if those efforts come at the expense of studies aimed at improving the assumptive and interpretive skills of engineers.

Since the artistic and synthetic aspects of engineering practice have remained fundamentally unaffected by developments in its scientific and analytical tools, the entire history of engineering continues to be relevant for the teaching and refinement of engineering thinking and judgment. The lessons learned from failures especially are as important today as they were in ancient times. This is not to say that the state of the art has not advanced, for it incontrovertibly has. But the errors in judgment made by a Roman engineer can be and have been repeated in modern times. The mistaken assumptions made by a Renaissance engineer can be and have been repeated in twentieth-century designs. Errors and mistakes of the past have been recorded by Vitruvius, Galileo, and others, and thus the design lessons in classic works are relevant to today's engineering – even if the mathematics has long been superseded.

Although the computer has displaced the slide rule and steel and concrete have displaced timber and stone, nothing has replaced the fact that an engineer must still sketch his or her concept before he or she can calculate its strength and cost. And engineers must still imagine how their structures can fail if they are to calculate factors of safety and reliability. Although science and mathematics have advanced, the laws of nature have not changed, and the same force of the wind that destroyed the decks of numerous early-nineteenth-century suspension bridges destroyed the Tacoma Narrows Bridge in 1940. It is as if all the analytical advances in suspension bridge design were for naught because the consequences of what proved to be the critical failure mode had not been analyzed.

Today the value of lessons learned from failures is widely recognized. But collectively very old as well as very recent failures hold information well beyond their obvious caveats. Rather than view the failure of the Tacoma Narrows Bridge, for example, as a lesson only for designers of suspension bridges with flexible decks, we can look beyond its most obvious lesson to seek what lessons it holds for the *process* of engineering design generally. By generalizing about classes of familiar and not-so-familiar failures, engineers can learn and teach what *kinds* of errors and mistakes of engineering thinking and judgment have been repeated throughout history.

There was a time when history of engineering was more widely studied by engineering students than it is now. Pressures to reduce overall degree requirements coupled with the desire for more and more advanced courses in mathematics and engineering analysis won out. History of engineering courses were generally viewed as a cultural luxury, perhaps helpful for giving students a professional pride but not important enough to require in the curriculum. Since such courses did tend to be celebrations of engineering success, sometimes filled with highly debatable minutiae about the construction of pyramids or cathedrals, it was hard to see how they related to designing space satellites or suspension bridges that did not fall down.

Today there are indications that the lessons of engineering history are sorely missed. In an open discussion at a joint meeting in 1976 of the British Institution of Mechanical Engineers and the Newcomen Society for the Study of the History of Engineering and Technology, it was proposed that in "rapidly advancing industries, ahead of textbooks, it is useful to find out how earlier engineers had coped with their problems" (Newcomen Society, 1976). Furthermore, it was felt that "the transfer of ideas from one field to another . . . is a concomitant of the study of engineering history." A more recent conference on the matter brought historians and engineers together, and their resulting joint statement of opportunity presents a strong case for giving "serious and active attention to how instruction in the history of engineering can be implemented in the revitalization of engineering education" (Vincenti and McGinn, 1992).

The history of engineering is relevant to even the highest of high technology, if it is understood that the proper focus of study is engineering thinking about the role of failure in achieving success, and not a mere recitation of engineering achievements (see Petroski, 1992e). History of engineering, especially in a form containing case studies of failures grouped under unifying themes relating to various aspects of the engineering method, would be an invaluable addition to the modern curriculum. And if some of these courses could be taught by reflective engineers who have experienced engineering first-hand, then students could not only learn something of the culture of their profession, they could come to learn from the masters how to develop engineering judgment. Furthermore, such courses could get students started on a lifelong path to reading engineering history on their own, discovering or recognizing new case studies, and perhaps writing them up for other engineers to read. In this way, some great lost lessons of engineering could be rediscovered and engineering thinking and judgment sharpened rather than dulled in future engineers.

Bibliography

Adams, Henry (1918). *The Education of Henry Adams: An Autobiography*. Boston: Houghton Mifflin.

Addis, William (1990). *Structural Engineering: The Nature of Theory and Design*. Chichester, West Sussex: Ellis Horwood.

Alexander, Christopher (1964). *Notes on the Synthesis of Form*. Cambridge, Mass.: Harvard University Press.

Allen, D. E. (1984). "Structural Failures Due to Human Error – What Research to Do?" In *Risk, Structural Engineering and Human Error*. Edited by M. Grigoriu. Waterloo, Ontario: University of Waterloo Press. Pp. 127–137.

American Society of Civil Engineers (1988). *Quality in the Constructed Project: A Guideline for Owners, Designers and Contractors*. Preliminary Edition for Trial Use and Comment. Vol. 1. New York: ASCE.

American Society for Engineering Education [1986]. *Engineering Case Library: A Catalog of Cases Available from ASEE*. Washington, D.C.: ASEE.

Ammann, O. H. (1933). "George Washington Bridge: General Conception and Development of Design." *Transactions of the American Society of Civil Engineers 97*, 1–65.

Anonymous (1989). "Is Ship Design All at Sea?" *The Engineer*, February 16, pp. 24–25.

(1992a). "New Cable-Stayed Bridge Will Span the Mississippi." *Civil Engineering*, February, pp. 16, 18.

(1992b). "Cable-Stayed Bridge Fails in Korea." *ENR*, August 10, pp. 7–8.

Aristotle (fourth century B.C.E.). *Minor Works.* Translated by W. S. Hett. Cambridge, Mass.: Harvard University Press, 1980.

Baker, B. (1873). *Long-Span Railway Bridges.* London: E. & F. N. Spon.

Beckett, Derrick (1980). *Brunel's Britain.* Newton Abbot, Devon: David & Charles.
(1984). *Stephensons' Britain.* Newton Abbot, Devon: David & Charles.

Bell, T. E., ed. (1989). "Special Report: Managing Risk in Large Complex Systems." *IEEE Spectrum,* June, pp. 22–51.

Bell, Trudy, and Karl Esch (1987). "The Fatal Flaw in Flight 51-L." *IEEE Spectrum,* February, pp. 36–51.

Benvenuto, Eduardo (1991). *An Introduction to the History of Structural Mechanics.* Two parts. New York: Springer Verlag.

Billah, K. Y., and R. H. Scanlan (1991). "Resonance, Tacoma Narrows Bridge Failure, and Undergraduate Physics Textbooks." *American Journal of Physics 59,* 118–124.

Billington, David P. (1977). "History and Aesthetics in Suspension Bridges." *Journal of the Structural Division, Proceedings of the American Society of Civil Engineers 103,* 1655–1672. See also discussions in vol. 104 by Herbert Rothman, pp. 246–249; George Schoepfer, pp. 378–379; Peter G. Buckland, pp. 379–380; Hasan I. A. Hegab, p. 619; Harold Samelson, pp. 732–733; Edward Cohen and Frank T. Stahl, pp. 1027–1030; Blair Birdsall, pp. 1030–1035; and John Paul Hartman, pp. 1174–1176. Billington's response is in vol. 105, pp. 671–687.
(1980). "Wilhelm Ritter: Teacher of Maillart and Ammann." *Journal of the Structural Division, Proceedings of the American Society of Civil Engineers 106,* 1103–1116.
(1983). *The Tower and the Bridge: The New Art of Structural Engineering.* Princeton, N.J.: Princeton University Press.

Billington, David P., and Aly Nazmy (1990). "History and Aesthetics of Cable-Stayed Bridges." *Journal of Structural Engineering 117,* 3103–3134.

Bishop, John George (1897). *The Brighton Chain Pier: In Memoriam. Its History from 1823 to 1896, with a Biographical Notice of Sir Samuel Brown, its Designer and Constructor, and An Appendix (Legal Documents).* Brighton: J. G. Bishop.

Blockley, D. I. (1980). *The Nature of Structural Design and Safety.* Chichester, West Sussex: Ellis Horwood.

Blockley, D. I., and J. R. Henderson (1980). "Structural Failures and the Growth of Engineering Knowledge." *Proceedings of the Institution of Civil Engineers 68,* Part 1, 719–728.

Boyd, G. M., ed. (1970). *Brittle Fracture in Steel Structures.* London: Butterworths.

Brown, Colin B. (1986). "Incomplete Design Paradigms." In *Modelling Human Er-*

ror in Structural Design and Construction. Edited by A. S Nowak. New York: American Society of Civil Engineers. Pp. 8–12.

Brown, Colin B., and Xiaochin Yin (1988). "Errors in Structural Engineering." *Journal of Structural Engineering 114*, 2575–2593.

Browne, Malcom W. (1983). "Disaster on I-95." *Discover,* September, pp. 14–15, 18–20, 22.

Bruneau, M. (1992). "Evaluation of System-Reliability Methods for Cable-Stayed Bridge Design." *Journal of Structural Engineering 118*, 1106–1120.

Brunel, Isambard (1870). *The Life of Isambard Kingdom Brunel, Civil Engineer.* London: Longmans, Green.

Burford, A. (1960). "Heavy Transport in Classical Antiquity." *The Economic History Review 13* (Second Series), 1–18.

Byrd, Ty (1992). "Pride of Place." *New Civil Engineer,* May 7, pp. 22–26.

Campbell, Robert (1988). "Learning from the Hancock." *Architecture,* March, pp. 68–75.

Canada, Department of Railways and Canals (1919). *The Quebec Bridge over the St. Lawrence River, near the City of Quebec, on the line of the Canadian National Railways.* Report of the Government Board of Engineers. Ottawa.

Clark, Edwin (1850). *The Britannia and Conway Tubular Bridges: With General Inquiries on Beams and on the Properties of Materials Used in Construction.* London: Day and Son.

Condit, Carl W. (1982). *American Building: Materials and Techniques from the First Colonial Settlements to the Present.* Second Edition. Chicago: University of Chicago Press.

[Cooper, Theodore] (1907). "On the Quebec Bridge and Its Failure." *Engineering News,* October 31, pp. 473–477.

Coulton, J. J. (1977). *Ancient Greek Architects at Work: Problems of Structure and Design.* Ithaca, N.Y.: Cornell University Press.

Cowan, Henry J. (1977). *The Master Builders: A History of Structural and Environmental Design from Ancient Egypt to the Nineteenth Century.* New York: John Wiley & Sons.

Cowin, S. C. (1983). "A Note on Broken Pencil Points." *Journal of Applied Mechanics 50*, 453–454.

Cronquist, D. (1979). "Broken-off Pencil Points." *Journal of Applied Physics 47*, 653–655.

Cross, N., ed. (1984). *Developments in Design Methodology.* Chichester, West Sussex: John Wiley & Sons.

Dana, Allston, Askel Andersen, and George M. Rapp (1933). "George Washington

Bridge: Design of Superstructure." *Transactions of the American Society of Civil Engineers 97*, 97–163.

Dibner, Bern (1950). *Moving the Obelisks: A Chapter in Engineering History in Which the Vatican Obelisk in Rome in 1586 Was Moved by Muscle Power, and a Study of More Recent Similar Moves*. New York: Burndy Library.

Dixon, J. R. (1988). "On Research Methodology Towards a Scientific Theory of Engineering Design." *Artificial Intelligence for Engineering Design, Analysis, and Manufacturing 1*, 145–157.

Dixon, J. R., and M. K. Simmons (1984). "Expert Systems for Engineering Design: Standard V-Belt Design as an Example of the Design-Evaluate-Redesign Architecture." *Proceedings of the ASME Computers in Engineering Conference* (Las Vegas), Vol. 1, pp. 332–337.

Dixon, J. R., et al. (1988). "A Proposed Taxonomy of Mechanical Design Problems." *Proceedings of the ASME Computers in Engineering Conference* (San Francisco), Vol. 1, pp. 41–46.

Drake, Stillman (1978). *Galileo at Work: His Scientific Biography*. Chicago: University of Chicago Press.

Ellingwood, B. (1987). "Design and Construction Error Effects on Structural Reliability." *Journal of Structural Engineering 113*, 409–423.

Emmerson, George S. (1977). *John Scott Russell: A Great Victorian Engineer and Naval Architect*. London: John Murray.

Fairbairn, William (1849). *An Account of the Construction of the Britannia and Conway Tubular Bridges. . . .* London: John Weal.

(1865). *Treatise on Iron Ship Building: Its History and Progress*. London: Longmans, Green.

(1877). *The Life of Sir William Fairbairn, Bart*. Edited by W. Pole. London: Longmans, Green.

Farquharson, F. B. (1949). *Aerodynamic Stability of Suspension Bridges, with Special Reference to the Tacoma Narrows Bridge. Part 1: Investigations Prior to October, 1941*. Report, Structural Research Laboratory, University of Washington.

Ferguson, E. S. (1977). "The Mind's Eye: Nonverbal Thought in Technology." *Science 197*, 827–836.

(1992). *Engineering and the Mind's Eye*. Cambridge, Mass.: MIT Press.

Finch, J. K. (1941). "Wind Failures of Suspension Bridges: Or, Evolution and Decay of the Stiffening Truss." *Engineering News-Record*, March 13, pp. 74–79. See also letter, March 27, p. 43.

(1978). *Engineering Classics*. Edited by Neal FitzSimons. Kensington, Md.: Cedar Press.

Fisher, J. W. (1984). *Fatigue and Fracture in Steel Bridges: Case Studies.* New York: John Wiley & Sons.

Fitchen, John (1986). *Building Construction before Mechanization.* Cambridge, Mass.: MIT Press.

FitzSimons, Neal (1988). "Bridge Failures: Errors Ignored." *Proceedings of the 2nd Historic Bridges Conference* (Columbus, Ohio), March 11, pp. 77–90.

Fowler, D. (1988). "The Shape of Bridges to Come." *New Civil Engineer,* June 30, pp. 31–32.

Fraczek, J. (1979). "ACI Survey of Concrete Structure Errors." *Concrete International,* December, pp. 14–20.

French, M. J. (1988). *Invention and Evolution: Design in Nature and Engineering.* Cambridge University Press.

Fung, Y. C. (1977). *A First Course in Continuum Mechanics.* Englewood Cliffs, N.J.: Prentice-Hall.

Galileo (1638). *Dialogues Concerning Two New Sciences.* Translated by H. Crew and A. de Salvio, 1914. New York: Dover, [1954].

Gasparini, D. A., and Melissa Fields (1993). "Collapse of Ashtabula Bridge on December 29, 1878." *Journal of Performance of Constructed Facilities 7,* 109–125.

Glegg, G. L. (1973). *The Science of Design.* Cambridge University Press.

(1981). *The Development of Design.* Cambridge University Press.

Gordon, J. E. (1978). *Structures: Or Why Things Don't Fall Down.* New York: Da Capo Press.

(1988). *The Science of Structures and Materials.* New York: Scientific American Library.

Great Britain (1849). *Report of the Commissioners Appointed to Inquire into the Application of Iron to Railway Structures.* London: William Clowes and Sons.

Great Britain, Department of the Environment (1973). *Inquiry into the Basis of Design and Method of Erection of Steel Box-Girder Bridges.* Committee Report. London: Her Majesty's Stationery Office.

Great Britain, Ministry of Housing and Local Government (1968). *Report of the Inquiry into the Collapse of Flats at Ronan Point, Canning Town.* London: Her Majesty's Stationery Office.

Great Britain, Navy Department, Advisory Committee on Structural Steels (1970). *Brittle Fracture in Steel Structures.* London: Butterworths.

Hall, A. Rupert (1978). "On Knowing, and Knowing How to. . . . " In *History of Technology. Third Annual Volume, 1978.* Edited by A. Rupert Hall and Norman Smith. London: Mansell. Pp. 91–103.

Hammond, K. J. (1986). "Learning to Anticipate and Avoid Planning Problems

Through the Explanation of Failure." *Proceedings of the Fifth National Conference on Artificial Intelligence* (AAAI-86, Philadelphia), pp. 556–560.

Hauser, R. (1979). "Lessons from European Failures." *Concrete International*, December, pp. 21–25.

Hersey, J. (1957). *A Single Pebble*. New York: Alfred A. Knopf.

Holgate, Alan (1986). *The Art in Structural Design: An Introduction and Sourcebook*. Oxford: Clarendon Press.

Hopkins, H. J. (1970). *A Span of Bridges: An Illustrated History*. New York: Praeger.

Hunt, Robert, ed. (1851). *Hunt's Hand-Book to the Official Catalogues: An Explanatory Guide to the Natural Productions and Manufactures of the Great Exhibition of the Industry of All Nations, 1851*. London.

Ingles, O. G. (1979). "Human Factors and Error in Civil Engineering." *Proceedings of the Third International Conference on Applications of Statistics and Probability in Soil and Structural Engineering* (Sydney), pp. 402–417.

Ingles, O. G., and G. Nawar (1983). "Evaluation of Engineering Practice in Australia." *Proceedings of the IABSE Workshop* (Rigi), pp. 111–116.

Institution of Civil Engineers (1973). *Steel Box Girder Bridges: Proceedings of the International Conference*. London: ICE.

Institution of Structural Engineers (1969). *Aims of Structural Design*. London: IStructE.

Ironbridge Gorge Museum Trust (1979). *The Iron Bridge: A Short History of the First Iron Bridge in the World*. Coalbrookdale, Shropshire: Ironbridge Gorge Museum Trust.

Johnston, Bruce G. (1965). "An Awareness of Reality." *Civil Engineering*, January, pp. 58–59.

Kaminetzky, Dov (1991). *Design and Construction Failures: Lessons from Forensic Investigations*. New York: McGraw-Hill.

Knoll, F. (1982). "Human Error in the Building Process – A Research Proposal." *IABSE Journal J-17/82*, 35–50.

(1984). "Modelling Gross Errors – Some Practical Considerations." In *Risk, Structural Engineering and Human Error*. Edited by M. Grigoriu. Waterloo, Ontario: University of Waterloo Press. Pp. 139–161.

(1986). "Checking Techniques." In *Modelling Human Error in Structural Design and Construction*. Edited by A. S. Nowak. New York: American Society of Civil Engineers. Pp. 26–42.

Koerte, Arnold (1992). *Two Railway Bridges of an Era: Firth of Forth and Firth of Tay*. Basel: Birkhäuser.

Kuesel, T. R. (1990). "Whatever Happened to Long-Term Bridge Design?" *Civil Engineering*, February, pp. 57–60.

Kuhn, Thomas S. (1962). *The Structure of Scientific Revolutions.* Chicago: University of Chicago Press.

Landels, J. G. (1978). *Engineering in the Ancient World.* Berkeley: University of California Press.

Larsen, Egon. (1969). *A History of Invention.* London: J. M. Dent & Sons.

Leonhardt, F. (1984). *Bridges: Aesthetics and Design.* Cambridge, Mass.: MIT Press.

Levy, Matthys, and Mario Salvadori (1992). *Why Buildings Fall Down: How Structures Fail.* New York: Norton.

Lind, N. C. (1983). "Models of Human Error in Structural Reliability." *Structural Safety 1*, 167–175.

(1986). "Control of Human Error in Structures." In *Modelling Human Error in Structural Design and Construction.* Edited by A. S. Nowak. New York: American Society of Civil Engineers. Pp. 122 127.

Lindenthal, Gustav (1922). "The Continuous Truss Bridge over the Ohio River at Sciotoville, Ohio, of the Chesapeake and Ohio Northern Railway," with appended discussion. *Transactions of the American Society of Civil Engineers 85*, 910–975.

Mainstone, R. (1977). "The Uses of History." *Architectural Science Review*, June, pp. 30–34.

Mark, Hans (1987). *The Space Station: A Personal Journey.* Durham, N.C.: Duke University Press.

Marlin, William (1977). "Some Reflections on the John Hancock Tower." *Architectural Record*, June, pp. 117–126.

Marshall, R. D., et al. (1982). *Investigation of the Kansas City Hyatt Regency Walkways Collapse.* U.S. Department of Commerce, National Bureau of Standards, Report NBSIR 82-2465.

Matousek, M. (1983). "Control Measures and Their Application." *Proceedings of the IABSE Workshop* (Rigi), pp. 161–164.

McCullough, David (1972). *The Great Bridge.* New York: Simon and Schuster.

Melchers, R. E. (1976). *Studies of Civil Engineering Failures: A Review and Classification.* Monash University, Civil Engineering Research Report No. 6/1976.

(1984). *Human Error in Structural Reliability – II: Review of Mathematical Models.* Monash University, Civil Engineering Research Report No. 3/1984.

(1987). *Structural Reliability: Analysis and Prediction.* Chichester, West Sussex: Ellis Horwood.

Melchers, R. E., and M. V. Harrington (1984). *Human Error in Structural Reliability – I: Investigation of Typical Design Tasks.* Monash University, Civil Engineering Research Report No. 2/1984.

Melchers, R. E., and M. E. Stewart (1985). "Data-Based Models for Human Error in Design." *Proceedings of the Fourth International Conference on Structural Safety and Reliability*, pp. II-51–II-60.

Melchers, R. E., M. J. Baker, and F. Moses (1983). "Evaluation of Experience." *Proceedings of the IABSE Workshop* (Rigi), pp. 21–38.

Moore, W. P., Jr. (1984). "Structural Strength and Safety – The Profession at a Crossroad." *Concrete International*, October, pp. 42–45.

National Aeronautics and Space Administration (1990). *The Hubble Space Telescope Optical Systems Failure Report*. NASA TM-103443 (November).

Navier, C. L. M. H. (1823). *Rapport à Monsieur Becquey . . . ; et Mémoire sur les ponts suspendus*. Paris.

Nessim, M. A., and I. J. Jordaan (1985). "Models for Human Error in Structural Reliability." *Journal of Structural Engineering 111*, 1358–1376.

Newcomen Society (1976). "The Value of Engineering History to the Engineer: An Open Discussion." *Transactions of the Newcomen Society 47*, 225–226.

Norman, Donald A. (1988). *The Psychology of Everyday Things*. New York: Doubleday. Reprinted in paperback as *The Design of Everyday Things*. New York: Doubleday, 1990.

Nowak, A. S., ed. (1979). "Effect of Human Error on Structural Safety." *ACI Journal 76*, 959–972.

   (1986). *Modelling Human Error in Structural Design and Construction: Proceedings of a Workshop*. New York: American Society of Civil Engineers.

Nowak, A. S., and R. I. Carr (1985a). "Classification of Human Errors." In *Structural Safety Studies*. Edited by J. T. P. Yao et al. New York: American Society of Civil Engineers. Pp. 1–10.

   (1985b). "Sensitivity Analysis for Structural Errors." *Journal of Structural Engineering 111*, 1734–1746.

Nowak, A. S., and S. W. Tabsh (1988). "Modelling Human Error in Structural Design." *Forensic Engineering 1*, 233–141.

O'Neill, B. (1991). "Bridge Design Stretched to the Limits." *New Scientist*, October 26, pp. 36–43.

Pannell, J. P. M. (1977). *Man the Builder: An Illustrated History of Civil Engineering*. New York: Crescent Books.

Pasley, C. W. (1842). "Description of the State of the Suspension Bridge at Montrose, After It Had Been Rendered Impassable by the Hurricane of the 11th of October, 1838; with Remarks on the Construction of that and other Suspension Bridges, in Reference to the Action of Violent Gales of Wind." *Transactions of the Institution of Civil Engineers 3*, 219–227.

Paxton, R. A. (1979). "Menai Bridge 1818–26: Evolution of Design." *Transactions of the Newcomen Society 49*, 27–110.

Paxton, R. A., ed. (1990). *100 Years of the Forth Bridge*. London: Thomas Telford.

Pechter, Kerry (1989). "Why the Holes in R & D?" *Across the Board*, September, pp. 32–36.

Peck, Ralph B. (1969). "A Man of Judgement." Second R. P. Davis Lecture on the Practice of Engineering. West Virginia University.

——— (1981). "Where Has All the Judgement Gone?" Norges Geotekniske Institutt, *Publikasjon* no. 134.

Penfold, Alastair, ed. (1980). *Thomas Telford: Engineer*. Proceedings of a Seminar . . . , Ironbridge, April 1979. London: Thomas Telford.

Petroski, Henry (1985). *To Engineer Is Human: The Role of Failure in Successful Design*. New York: St. Martin's Press.

——— (1987a). "Design as Obviating Failure." In *Plenary and Interdisciplinary Lectures of the 1987 International Congress on Planning and Design Theory*. Edited by Gerald Nadler. New York: American Society of Mechanical Engineers. Pp. 49–53.

——— (1987b). "On the Fracture of Pencil Points." *Journal of Applied Mechanics 54*, 730–733.

——— (1988). "Engineering Failures and Professional Responsibility." *Forensic Engineering 1*, 149–152.

——— (1989a). "Design Errors as Causes of Failures: Their Reduction Through the Use of Case Histories." *Proceedings of the International Conference on Case Histories in Structural Failures* (Singapore), pp. D-22–D-30.

——— (1989b). "Failure as a Unifying Theme in Design." *Design Studies 10*, 214–218.

——— (1990). *The Pencil: A History of Design and Circumstance*. New York: Alfred A. Knopf.

——— (1991a). "On the Backs of Envelopes." *American Scientist*, January–February, pp. 15–17.

——— (1991b). "Good Drawings and Bad Dreams." *American Scientist*, March–April, pp. 104–107.

——— (1991c). "In Context." *American Scientist*, May–June, pp. 202–204.

——— (1991d). "Still Twisting." *American Scientist*, September–October, pp. 398–401.

——— (1992a). "Isambard Kingdom Brunel." *American Scientist*, January–February, pp. 15–19.

——— (1992b). "Making Sure." *American Scientist*, March–April, pp. 121–124.

——— (1992c). "The Britannia Tubular Bridge." *American Scientist*, May–June, pp. 220–224.

——— (1992d). "The Evolution of Artifacts." *American Scientist*, September–October, pp. 416–420.

——— (1992e). "History and Failure." *American Scientist*, November–December, pp. 523–526.

(1992f). *The Evolution of Useful Things*. New York: Alfred A. Knopf.

(1993). "Predicting Disaster." *American Scientist,* March–April, pp. 110–113.

Petroski, Henry, ed. (1992g). "Design in a Human Context." *Research in Engineering Design,* vol. 4, no. 1.

Picon, Antoine (1988). "Navier and the Introduction of Suspension Bridges in France." *Construction History 4,* 21–34.

Podolny, Walter, Jr. (1992). "Cable-stayed Bridges: Future Developments." Presented at the American Society of Civil Engineers Annual Convention, New York, September 13–17.

Potter, I. M. D. (1981). "Avoidance of Error in Design Processes." In *Structural Failures in Buildings*. London: Institution of Structural Engineers.

Prebble, John (1975). *The High Girders*. London: Book Club Associates.

Provis, W. A. (1842). "Observations on the Effects of Wind on the Suspension Bridge over the Menai Strait, more Especially as Relates to the Injuries Sustained by the Roadways during the Storm of January, 1839; together with brief Notices of Various Suggestions for Repairing the Structure." *Transactions of the Institution of Civil Engineers 3,* 357–370.

Pugsley, Sir Alfred (1968). *The Theory of Suspension Bridges*. London: Edward Arnold.

Pugsley, Sir Alfred, ed. (1976). *The Works of Isambard Kingdom Brunel: An Engineering Appreciation*. London: Institution of Civil Engineers.

Pugsley, A., R. J. Mainstone, and R. J. M. Sutherland (1974). "The Relevance of History." *The Structural Engineer 52,* 441–445. (See also the discussion in vol. 53, pp. 387–398.)

Rankine, W. J. M. (1856). *Introductory Lecture on the Harmony of Theory and Practice in Mechanics*. London: Richard Griffin.

Rendel, J. M. (1841). "Memoir on the Montrose Suspension Bridge." *Minutes of Proceedings,* Institution of Civil Engineers, vol. 1, pp. 122–127.

Richards, K. G. (1971). *Brittle Fracture of Welded Structures*. Abington, Cambridgeshire: The Welding Institute.

Roddis, W. M. Kim (1993). "Structural Failures and Engineering Ethics." *Journal of Structural Engineering 119,* 1539–1555.

Roebling, John A. (1841). "Remarks on Suspension Bridges, and on the Comparative Merits of Cable and Chain Bridges." *American Railroad Journal, and Mechanics' Magazine 6* (n.s.), 193–196.

(1855). *Final Report to the Presidents and Directors of the Niagara Falls Suspension and Niagara Falls International Bridge Companies*. Rochester, N.Y.: Lee, Mann.

(1867). *Annual Report of the President and Directors to the Stockholders of the*

Covington and Cincinnati Bridge Company for the Year Ending Feb. 28th, 1867.
Trenton, N.J.: Murphy & Bechtel.

——— (1870). *Report to the President and Directors of the New York Bridge Company, on the Proposed East River Bridge.* Brooklyn: Daily Eagle Print.

[Roebling, W. A.] (1877). *Report of the Chief Engineer of the New York & Brooklyn Bridge, January 1, 1877.* Brooklyn: Eagle Print.

Rolt, L. T. C. (1970). *Isambard Kingdom Brunel.* Harmondsworth, Middlesex: Penguin Books.

Rosenberg, N., and W. G. Vincenti (1978). *The Britannia Bridge: The Generation and Diffusion of Technological Knowledge.* Cambridge, Mass.: MIT Press.

Rosenthal, Andrew (1989). "Trident 2 Failures Laid to Early Success." *The New York Times*, August 18.

Ross, S. S. (1984). *Construction Disasters: Design Failures, Causes, and Prevention.* New York: McGraw-Hill.

Russell, J. S. (1839). "On the Vibration of Suspension Bridges and Other Structures; and the Means of Preventing Injury from this Cause." *Transactions, Royal Scottish Society of Arts 1*, 304–314.

——— (1864). *The Modern System of Naval Architecture.* London: Day and Son, 1864–65.

Santamarina, J. C., and J. L. Chameau (1989). "Limitations in Decision Making and System Performance." *Journal of Performance of Constructed Facilities 3*, 78–86.

Shipway, J. S. (1990). "The Forth Railway Bridge Centenary 1890–1990: Some Notes on Its Design." *Proceedings of the Institution of Civil Engineers 88*, Part 1, 1079–1107.

Shirley-Smith, Sir H. (1976). "Royal Albert Bridge, Saltash." In *The Works of Isambard Kingdom Brunel: An Engineering Appreciation.* Edited by Sir Alfred Pugsley. London: Institution of Civil Engineers.

Shute, Nevil (1954). *Slide-Rule: The Autobiography of an Engineer.* New York: Ballantine Books.

Sibly, P. G. (1977). "The Prediction of Structural Failure." Ph.D. thesis, University of London.

Sibly, P. G., and A. C. Walker (1977). "Structural Accidents and Their Causes." *Proceedings of the Institution of Civil Engineers 62*, 191–208.

Singh, C. J. (1976). *Bibliography of Structural Failures, 1850–1970.* Garston, England: Building Research Establishment.

Smiles, Samuel (1904). *Lives of the Engineers.* Five volumes. London: John Murray.

Spector, A., and D. Gifford (1986). "A Computer Science Perspective on Bridge Design." *Communications of the ACM 29*, 268–283.

Squires, A. M. (1986). *The Tender Ship: Governmental Management of Technological Change*. Boston: Birkhaüser.

Stauffer, L. A. (1987). "An Empirical Study on the Process of Mechanical Design." Ph.D. thesis, Oregon State University.

Stauffer, L. A., and D. G. Ullman (1988). "A Comparison of the Results of Empirical Studies into the Mechanical Design Process." *Design Studies 9*, 107–114.

Steinman, D. B. (1950). *The Builders of the Bridge: The Story of John Roebling and His Son*. New York: Harcourt, Brace.

Stewart, M. G. (1987). "Control of Human Errors in Engineering Design." Ph.D. thesis, University of Newcastle, New South Wales, Australia.

Stewart, M. G., and R. E. Melchers (1987). "Human Error in Structural Reliability – VI: Overview Checking." Report, Department of Civil Engineering and Surveying, University of Newcastle, New South Wales, Australia.

    (1989). "Checking Models in Structural Design." *Journal of Structural Engineering 115*, 1309–1324.

Straub, Hans (1949). *A History of Civil Engineering: An Outline from Ancient to Modern Times*. Translated by Erwin Rockwell, 1952. Cambridge, Mass.: MIT Press, 1964.

Sullivan, Walter (1986). "Divers Report No Hull Gash in the Titanic." *The New York Times*, July 31, pp. A1, A16.

Takena, K., et al. (1992). "Fatigue Resistance of Large-Diameter Cable for Cable-Stayed Bridges." *Journal of Structural Engineering 118*, 701–715.

Thomson, George H. (1888). "American Bridge Failures: Mechanical Pathology, Considered in Its Relation to Bridge Design." *Engineering*, September 14, pp. 252–253; September 21, p. 294.

Timoshenko, Stephen P. (1953). *History of Strength of Materials: With a Brief Account of the History of Theory of Elasticity and Theory of Structures*. New York: McGraw-Hill.

Todhunter, Isaac, and Karl Pearson (1886). *A History of the Theory of Elasticity and of the Strength of Materials from Galilei to Lord Kelvin*. Vol. 1. New York: Dover, 1960.

Tuchman, Janice L. (1987). "Man of the Year: John W. Fisher." *ENR,* February 19, pp. 40–50.

Turner, Roland, and Steven L. Goulden, eds. (1981). *Great Engineers and Pioneers in Technology*. Vol. 1. New York: St. Martin's Press.

Ullman, D. G., T. G. Dietterich, and L. A. Stauffer (1988). "A Model of the Mechanical Design Process Based on Empirical Data: A Summary." In *Artificial Intelligence in Engineering: Design*. Edited by J. S. Gero. Amsterdam: Elsevier. Pp. 193–215.

Ullman, D. G., L. A. Stauffer, and T. G. Dietterich (1987). "Toward Expert CAD." *Computers in Mechanical Engineering,* December, pp. 56–70.

U.S. House of Representatives, Committee on Science and Technology (1984). *Structural Failures in Public Facilities.* Report 98–621. Washington, D.C.: Government Printing Office.

U.S. Presidential Commission (1986). *Report to the President on the Space Shuttle Challenger Accident.* Washington, D.C.: Government Printing Office.

Vaughan, Adrian (1991). *Isambard Kingdom Brunel: Engineering Knight Errant.* London: John Murray.

Vincenti, Walter G. (1990). *What Engineers Know and How They Know It: Historical Studies in the Nature and Sources of Engineering Knowledge.* Baltimore: Johns Hopkins University Press.

Vincenti, Walter G., and Robert E. McGinn, co organizers (1992). "The History of Engineering for Engineering Students: A Challenge for Engineering Education." Report on a Conference. Stanford, Calif.: Program in Values, Technology, Science, and Society and School of Engineering, Stanford University.

Vitruvius (first century B.C.E.). *The Ten Books on Architecture.* Translated by M. H. Morgan, 1914. New York: Dover, 1960.

Waldron, K. J., and M. B. Waldron (1987). "A Retrospective Study of a Complex Mechanical System Design." Oakland, Calif.: NSF Workshop on Design Theory and Methodology.

Waldron, M. B., K. J. Waldron, and D. H. Owen (1988). "Use of Systematic Theory to Represent the Conceptual Mechanical Design Process." *Design Theory '88.* Preprints of the 1988 NSF Grantee Workshop on Design Theory and Methodology, Rensselaer Polytechnic Institute.

Walker, A. C. (1981). "Study and Analysis of the First 120 Failure Cases." In *Structural Failures in Buildings.* London: Institution of Structural Engineers. Pp. 75–83.

(1982). "History as a Predictor." *Proceedings of an IABSE (British Group) Colloquium on History of Structures* (Cambridge), pp. 75–83.

Walker, Jearl (1979). "The Amateur Scientist." *Scientific American,* February, pp. 158, 160, 162–166. See also November, pp. 202–204.

Wang, J., and H. C. Howard (1988). "Design-Dependent Knowledge for Structural Engineering Design." In *Artificial Intelligence in Engineering: Design.* Edited by J. S. Gero. Amsterdam: Elsevier. Pp. 267–277.

Watson, J. G. (1988). *The Civils: The Story of the Institution of Civil Engineers.* London: Thomas Telford.

Westhofen, W. (1890). "The Forth Bridge." *Engineering,* February 28, pp. 213–283.

Whyte, R. R., ed. (1975). *Engineering Progress Through Trouble*. London: Institution of Mechanical Engineers.

Williams, E. (1957). "Some Observations of Leonardo, Galileo, Mariotte and Others Relative to Size Effect." *Annals of Science 13:* 23–29.

Woodward, C. M. (1881). *A History of the St. Louis Bridge*. St. Louis: Polytechnic School, Washington University.

Zetlin, Lev (1988). "Compilation of Lectures Presented at Various National Conventions, Conferences, Seminars." West Palm Beach, Fla.: Zetlin-Argo Structural Investigations.

Zetlin, L., Associates (1978). *Report of the Engineering Investigation Concerning the Causes of Collapse of the Hartford Coliseum Space Truss Roof on January 18, 1978*. New York: Lev Zetlin Associates.

Index

Page numbers in *italics* refer to figures, captions, or tables.